THE TREATMENT AND MANAGEMENT OF URBAN SOLID WASTE

Samuel W. Bodman, III
James A. Freaney
Joseph J. Harrington
David H. Marks
Walter Niessen
Lawrence J. Partridge
Adel Sarofim

Edited by
DAVID GORDON WILSON
Massachusetts Institute of Technology

a TECHNOMIC® publication
TECHNOMIC Publishing Co., Inc.
265 W. State St., Westport, Conn. 06880

THE TREATMENT AND MANAGEMENT OF URBAN SOLID WASTE

265 W. State St., Westport, Conn. 06880

a TECHNOMIC® publication

Printed in U. S. A.

Library of Congress Card No. 70-182527
Standard Book No. 0-87762-077-6

TABLE OF CONTENTS

CHAPTER 1
HISTORY AND BACKGROUND OF SOLID-WASTE PROBLEMS

CHAPTER 2
SOLID-WASTE MANAGEMENT IN HOMES, APARTMENT AND OFFICE BUILDINGS

CHAPTER 3
URBAN REFUSE COLLECTION AND STREET CLEANING

CHAPTER 4
QUANTITATIVE ESTIMATES FOR SOLID-WASTE MANAGEMENT IN BOSTON

CHAPTER 5
OPERATIONS RESEARCH AND SOLID-WASTE COLLECTION

CHAPTER 6
LANDFILL

CHAPTER 7
MUNICIPAL INCINERATION

CHAPTER 8
RECLAMATION FROM MUNICIPAL SOLID WASTE

CHAPTER 9
COMPOSTING MUNICIPAL REFUSE

LIST OF FIGURES

LIST OF FIGURES

LIST OF TABLES

FOREWORD

The preparation of this book was suggested to the participants in a summer study on solid wastes, held at M.I.T. in 1968. The study attracted national attention, some of it favorable.

The original suggestion was for a revision of the summer-study report. However, each of the prospective authors wanted to make a fresh start rather than use as a basis a report that was inevitably written in rushed circumstances.

Furthermore, following the completion of the study, most participants immediately became deeply involved in various research programs, and were therefore anxious to incorporate the latest information. Accordingly, this book has taken more than two years to put together, because when academic people become involved in something they hold themselves virtually incommunicado for long periods.

This book is therefore completely different from the 1968 summer study. It contains new material by the same authors and, most importantly, new material by new authors. All the participants in the original study felt that they had little right to launch a work purporting to be a survey of the present field of solid-waste management from so narrow a base as M.I.T. We were fortunate already in having Mr. James A. Freaney as a participant in our summer study, bringing with him wide range of practical experience as the founder and manager of an organization involved in refuse collection and street cleaning, among several other activities, in the City of Boston.

Additionally, we were joined by Mr. Walter Niessen of Bolt, Beranek and Newman, Inc. Walter Niessen was in charge of a broad survey of the air-pollution implications of incineration and associated solid-waste practices for the National Air-Pollution-Control Administration, U.S. Department of Health, Education and Welfare. He collaborated with Adel Sarofim in producing the section on incineration in this present volume.

The area of handling, collection, and transportation has received valuable contributions from several new sources. One is David H. Marks, who recently joined the M.I.T. faculty, having been previously on the staff of the Federal Water-Pollution Administration of the Department of the Interior. He completed his doctorate at the Johns Hopkins University, working in a group that has become well known for its contributions to the application of operations-research methods to the management of solid-waste collection systems with emphasis on siting and on collection routing and scheduling.

As our last additional contribution, we were delighted to be able to include a case study on refuse-collection and street-cleaning practices in Boston from the Department of Environmental-Health Science, Harvard School of Public Health. Lawrence J. Partridge has been engaged in a study of various aspects of the Boston municipal public-health scene, is completing his doctoral dissertation under the supervision of Joseph J. Harrington, and in the course of this work has collaborated with James Freaney. The group at Harvard has a continuous history of strong activity in sanitary engineering and has recently been involved in a broad study of urban solid-wastes problems and practises. Support came from the Ford Foundation and the National Science Foundation.

The other parts of this book are written by original contributors to the summer-study, but in general owe much to research programs and other activities which have occurred since the study was completed. Samuel Bodman has been studying financial and contractual factors both in his former role of technical director of American Research and Development and as one of a group of former summer-study participants which has given a number of one- and two-day seminars since that time. Adel Sarofim has been supervising research on incineration, funded by the U.S. Public Health Service, Bureau of Solid-Waste Management (now the Environmental Protection Agency). Another bureau-funded research program has been supervised by David Wilson in the area of sorting and slavage, and the chapter on this subject derived from papers presented at the Houston Second Annual Industrial Solid-Waste Management Conference and the Second Annual Institute of the Technical Guidance Center, University of Massachusetts.

The editor acknowledges with pleasure and gratitude the willing cooperation and forebearance of the contributors to this book; the smooth and efficient relationship with Technomic Publishing Co., Inc. and Mrs. E. K. Moulketis; and the cheerful helpfulness of Miss Carol Grimes, who typed the final version and some which went before.

CHAPTER 1

HISTORY AND BACKGROUND OF SOLID-WASTE PROBLEMS

INTRODUCTION

The treatment of solid wastes in the United States has traditionally been given a very low priority on almost every level – governmental, planning, research, corporate and individual. Attention has turned to problems of solid wastes only, in general, when a local crisis has developed – the barrels are full; the dump is full; the incinerator pollutes and is to be shut down. Quite suddenly, it seems that many local crises have coalesced into a national crisis. There are cynics who protest that the situation is no worse than it has been for many years past. The rising national concern, they maintain, is a fad. It will pass away as have – by definition – other fads. Attention will shift, capriciously, to another aspect of national life.

The cynics may be right. But after the peak of concern has passed, the public attitude will not return to its previous torpor. The process of reeducation will have affected everyone, and it will be almost all beneficial. There is a rising demand for better solutions, where before apathy had been the rule. There is even a readiness to appropriate money for improved methods of treatment and for research. The Solid-Waste Act of 1965 has sparked developments over a wide spectrum. The activity is now so intense that an intelligent layman, and even a professional, may well be confused by the stream of news releases, technical papers, patent disclosures and forecasts about new methods of waste treatment. Many of the claims promise, along with many other benefits, substantially reduced costs, or even of profits, from solid-waste treatment.

The flow of developments has had an unfortunate side effect. A new form of community inaction has been brought about. It is a response to the desire to wait for another year or two until some of these promised advantages become realities. For there are as yet extremely few examples of new techniques of solid-waste treatment in successful operation. Is it worth waiting? Are some of the methods presently being developed sufficiently reliable in concept that they can be ordered without a full-scale demonstration trial? If present-day technology is to be relied upon, what are the best of current practices? These questions, which are being put by city officials and concerned citizens, are of the type this book attempts to answer.

Simultaneously we endeavor to point out the areas where research, or the development of new products or services, or the assembly of a new system from existing components, would bring benefits. We wish to give some guidance to those industrial and commercial organizations and research groups which may not have had previous experience in this field and may need to know what are the problem areas where progress is possible.

There are many areas where a comparatively small amount of planning and investment would bring disproportionate returns. Historically, solid wastes could be ignored to a much greater extent than liquid wastes, and when municipalities began to organize community action, the collection of refuse was the least popular and lowest paid of jobs.

In municipal departments responsible for refuse collection and disposal, innovative practices, or even good management, have usually been lacking. Commercial organizations which might have been able to satisfy requirements that might seem obvious have found that an absence of record keeping and simple measurements has made many people in authority unaware of their real needs. The private sector of the collection and disposal industry has, partly because of the inefficient nature of municipal solid-waste treatment, been composed principally of very small firms which have not been able to introduce innovative practices. Also, some have lacked the power to resist underworld influence in some parts of the country, and a stigma has penalized the whole industry.

SOLID WASTES IN BUILDINGS

The low status accorded to solid-waste treatment has resulted, somewhat surprisingly, in a general lack of sound and comprehensive ordinances and regulations governing the collection and storage of refuse in buildings. The result has been that in private homes from the cheapest to the most expensive available there are usually no facilities whatsoever included as part of the structure or the mechanical services, with the sole exception in some areas of a buried container for food wastes. What is even more surprising is that there are apartment houses and institutions being designed and erected in the 1970's which lack almost any provision for the handling, storage, and collection of solid wastes Yet as soon as buildings are handed over to the new owners or management, complaints regularly arise because of the problems of collecting refuse from individual offices or apartments and of bringing them to a central transfer location. It becomes difficult to keep litter from disfiguring many areas. One or more rooms are often designated as "rubbish rooms" and immediately become extreme fire risks. We have found in the architectural profession little concern to change practices and procedures to bring about automatic consideration of waste handling and storage at the planning stage of a new building. An aim of this book is to provide information about what is possible and what is needed within a single building.

When more than one building is planned, whether it be for a group of apartment buildings, a school campus, an industrial park, or even a new town, there are other measures which should be considered for the handling, storage, transportation, and disposition of solid wastes. This again is an area where highly beneficial approaches are possible if instituted at the planning stage, but which are almost impossible to incorporate later.

LEGISLATION

These examples illustrate a national attitude towards solid wastes which has until recently been marked by a lack of concern. Our attitudes are reflected in our laws, which in turn mold our actions. Presently we have a range of conflicting legislation, much of which has obviously been framed by people who in the past have not had a sound grasp of what is good for their own community and for the nation. For instance, there are laws which prevent refuse from one city being transported through another city, or which prevent collaboration among cities which might with advantage take a regional approach. The eventual necessity of recycling a large proportion of our solid wastes is acknowledged by most thinking people, but the degree of recycling which should be taking place at the present time is not attained because our legislation provides incentives for the use of raw materials and disincentives for salvage operations. We intend in this book to give some pointers to legislators and public officials towards desirable legislation and regulations.

AREAS NOT COVERED

Having reviewed the areas we wish to discuss, we need also to emphasize that we will not be attempting to deal with the vast and little-understood field of factory, farm, and forestry wastes. These are areas which, because of changing economics, are themselves changing fast. Surveys of the present status of solid-waste treatment in these fields have only recently been started.

HISTORY

One theory for man's loss of his fur is that when he adopted a mode of living that entailed keeping a fixed home base in contrast to the wandering habits of the rest of the ape family, he could be infested by fleas, the complicated life cycle of which depended in one phase on the presence of dirt and solid waste in the nest.[1] At some point man realized that he had to keep his home clean, and he accomplished this by throwing his waste outside to accumulate in the paths around. From archaeological investigations of past civilizations, it seems that this practice was so widespread that cities literally buried themselves in trash, sometimes having to rebuild repeatedly on top of the accumulation of past years.

The problem was so acute in some of the densely populated walled medieval cities that refuse had to be removed outside the community, rather than merely outside one's own dwelling. Laws were passed requiring that people who brought goods or produce into a city should carry solid wastes away with them for disposal in the countryside. Presumably this action was taken because of the nuisance of trash, rather than any realization of the connection between vermin and insects on the one hand and disease on the other.

Some interesting records about solid-waste treatment and regulation in London have been summarized by the Greater London Council in London's Refuse, from which the following passage is taken.

> From as early as 1297 there was a legal obligation on every householder in London to ensure that the pavement within the frontage of his tenement was kept clear. In 1354 an order was issued that filth deposited before houses was to be removed weekly. It was the duty of assistants to the Beadle of each ward known as rakers to ensure that all refuse was removed from the middle of the streets and from the fronts of houses. They had power to levy a charge on those householders who failed to remove their refuse.
>
> The refuse was raked together and loaded into tumbrils, drawn by two horses. London maintained 12 of these specially designed carts with 2 ft. 6 in. collapsible tailboards. A committee was elected in 1387 to discuss the provision of laystalls or places where rakers and carters might deposit what they had collected. A number of these laystalls were established in the city suburbs and on the banks of the River. Special days were appointed when refuse was to be put outside doors for the rakers to scoop up and trundle away to the laystalls. Londoners were instructed by an order of 1407 to keep their rubbish indoors until it could be carried away by the rakers. Refuse from the laystalls in the suburbs was sold to farmers and market gardeners; that from the riverside laystalls was taken downstream in boats to be dumped on the Essex marshes.
>
> Another method of disposal in medieval London was to throw refuse either into watercourses or the River itself. There were many official pronouncements against this practice.

[1] Desmond Morris, *The Naked Ape*, McGraw-Hill, 1967.

Edward III, after suffering from fumes and other abominable stenches arising from the river during a journey down the Thames in 1357, besought the Mayor and Sheriffs of London to forbid the throwing of rubbish or filth into the Thames, the Fleet or into the fosses around the walls of the city. An ordinance was directed in 1383 against those perons who had houses on the Walbrook and who, by throwing their refuse into the watercourse, had caused it to be stopped up to the great nuisance and damage of the city. In 1388 an Act was passed prohibiting the pollution of rivers, ditches and watercourses by deposits of filth and garbage and ordering that all such refuse should be carried away instead to appointed places before it became a source of nuisance.

The arrangements for refuse collection and disposal which had been established in London by the beginning of the fifteenth century remained essentially unchanged for some 500 years.

The authorities found it extremely difficult to enforce the regulations. In 1414 the Constables and Beadles of each of the City wards had to declare their willingness to pay informers to gather evidence against servants, ostlers and other offenders who cast rubbish and dirt into the streets. Such anti-social behavior is typified by William atte Wood who in 1421 was arraigned for making a great nuisance and discomfort to his neighbors by throwing out horrible filth on to the highway, the stench of which was so odious, that none of his neighbors could remain in their shops. . . .

At about this time the powers of the rakers or scavengers as they were commonly known by then were extended to include the clearing of large public spaces such as market squares and those portions of streets where houses were unoccupied and which previously had been sources of nuisance. Even so the King had cause to complain in 1628 that the ways in and about the city were very noisome and troublesome for passing because of the excessive quantities of filth lying in the street.

The Great Fire of London in 1666 had a great purifying effect and for some time the complaints about refuse in the streets ceased. By 1741, however, Lord Tyrconnel could describe the streets of London as abounding with such heaps of filth as a savage would look on with amazement. Gwynn in his Essay on Improvements of 1750 includes ordure and rubbish lying in the streets amongst the nuisances to be removed. Another visionary, Corbyn Morris, proposed two years later that the cleansing of the entire metropolitan area should be put under one uniform public management, and all the filth carried into lighters and conveyed by the Thames to proper distances in the country.

No action was taken, however, and the piles of refuse remained. One great heap at the bottom of Grays Inn Lane was not moved for a century until 1815 when the dust was extracted and sold to Russia to make brick for the rebuilding of Moscow after Napoleon's invasion. As recently as 1926, 41 brickfields were using dust from London's refuse to produce stock bricks, and some continue to do so to this day. . . .

It was some time, however, before the recommendation contained in Chadwick's General Report on the Sanitary Condition of the Laboring population of Great Britain of 1842 that public authorities undertake the removal of all refuse from habitations, streets and roads, was implemented. The modern system of refuse collection and disposal by the local authority derives from the Public Health Act of 1875. This Act made provision for the removal by the Sanitary Authority on appointed days of accumulations of refuse from premises. Each occupier was obliged to place the refuse in a movable receptacle – the first legal recognition of the dustbin.

Every Sanitary Authority was directed by the subsequent Public Health Act of 1891 to employ, or contract with, a sufficient number of scavengers to ensure the sweeping and the cleansing of the several streets within their district and the collection and removal of street and house refuse. The Authorities were given power to make by-laws with regard to these services. Arrangements for the disposal of refuse in the Greater London area were in the hands of some 90 local authorities until 1965.

One cannot help being both cheered and saddened by this encapsulated history: cheered at the realization that we have faced the solid-waste crisis for so long without being overwhelmed; and saddened that we have accomplished so little. We are still trying to achieve regionalization of refuse treatment; the strongest demand is still to take the trash away to a place distant from ourselves; regulations still seem seldom observed; and trash is still dumped into the street in many cities.

Indeed, the "modern system of refuse collection and disposal" instituted in 1875 has been changed little by modern technology. Men still accompany a wheeled vehicle and load it, usually by hand, after which the material is taken to be dumped or burned. The horses formerly used have been replaced by an internal-combustion engine which has not greatly increased the speed of collection in towns because of the presence of many other internal-combustion-engined vehicles. In fact, to some extent the productivity of the collection crew has fallen because while a horse can move on command, a motor vehicle has to have a driver, who usually does not take part in the collection process.

The major changes which have occurred in general refuse collection have largely been a consequence of the changes in the character of refuse. Formerly, houses were heated by fires, air-pollution ordinances were nonexistent, and accordingly anything combustible was burned. Before the use of tin cans became widespread, the incombustible output of solid wastes per person per day must have been very small indeed. In rural communities ashes were put on the land for their nutritive and conditioning value, but, in towns, ashes formed the bulk of the refuse. The introduction of the ash can ('dustbin' in the historical description of London's solid-waste treatment) a simplified form of containerization which we tend to take for granted, led to some improvement in conditions of collection.

Even when containerized, however, ashes were dirty to collect, and the next major technological advance was the development in Germany of several systems of so-called "dustless loading". Collection vehicles were equipped with automatic handling mechanisms which enclosed the mouth of each barrel, lifted, inverted, emptied, and cleaned it, and then returned it for refilling. This comparatively high level of sophistication has not been retained for collection under present conditions.

For as solid fuels were replaced by oil and gas for home heating, and the burning of paper and other trash was frowned upon or outlawed in more and more communities, and as we have become sufficiently educated and sufficiently affluent to afford more newspapers, magazines, and consumer goods, our domestic solid wastes have become more and more composed of paper and containers made of cardboard, wood, metal, glass, and plastics. The slow speed of operation of dustless-loading equipment has made it lose favor, and the increased bulk and decreasing density of solid wastes has brought about the last major change in collection technology, which is the compaction-type collection vehicle.

THE PRESENT SCENE

The "packer truck" forms part of what we shall define as the typical present method of urban refuse collection. We shall use this as a base case with which to compare alternatives. At the source – home, supermarket, apartment building, hospital, and so on, – loose refuse is containerized in trash barrels or larger containers. A compaction-type refuse-pickup truck of typically twenty-cubic-yard compacted capacity with a driver and two collectors empties the containers once a week and removes the trash to an incinerator or a dump or landfill. Across the United States, the bulk of refuse is still burned at the dump and covered only as a final resort when the dump is closed. Very little salvage of general refuse occurs. Items such as refrigerators and ranges, too large to be lifted by the collectors and accepted by the hopper at the rear of the collection vehicle, are sometimes the responsibility of the householder to dispose of and sometimes are collected by open pickup trucks in a special collection. In some communities, special collections are also made for yard refuse and such obvious public-health hazards as dead animals. Demolition refuse is, however, usually the responsibility of the contractor or householder to deal with on a private, commercial basis.

Another separate collection in which different vehicles and crews are employed is often undertaken once or twice a week for garbage. When this is done, a buried pail having a tightly covered lid is usually fitted outside. The accumulated food wastes are often cooked and used for pig food.

There are many variations of this composite picture of the typical treatment of solid wastes in a community. We shall examine the individual possibilities of variation in the remainder of this book.

CHAPTER 2

SOLID-WASTE MANAGEMENT IN HOMES, APARTMENT AND OFFICE BUILDINGS

SUMMARY

The problems and some promising solutions for the management of solid wastes in private homes, hotels, apartment buildings, hospitals, and supermarkets are discussed. Some of the better approaches available at present are compaction, recycling, and tube-transport systems. Developments which are either probable or desirable for the future are discussed.

INTRODUCTION

One major area which at present seems almost certain to be overlooked in the design of a building is the handling of solid wastes. Indeed, the handling of almost any materials is not given the attention it deserves. Architects of apartment buildings, for instance, seem to feel that if they make provisions for the movement of people, then the movement of goods of various kinds should be taken care of.

But just as the kitchens in modern hotels are no longer placed at the end of long corridors because labor is no longer cheap, and the comfort and convenience of employees are no longer a matter to be ignored, the movement of solid wastes must get more attention because labor costs and the ability to attract labor are strongly affected. Likewise, few people can afford servants in their homes and homeowners are very receptive to improved methods of handling refuse.

At present the best treatment of the problem of moving materials that can be hoped from the building designer is an intelligent layout of the interior space coupled possibly with freight elevators and perhaps refuse chutes. What may be desirable in large buildings of the future is a complete materials-handling system which will automatically convey virtually anything within certain size limits (for example, correspondence, or a tray set with a meal, or a bag of used diapers) from any one point to any other point in a building.

We are not at this stage yet; if we were, we would have to consider what special steps we would have to take to enable one handling system to be used for food in one direction and potentially disease-producing materials in another. Presently available approaches provide systems for waste handling which are completely separate from any other facilities.

THE PRINCIPAL PROBLEMS OF SOLID WASTES IN BUILDINGS

Solid waste is often described as a "nuisance". It is worth looking at the problem in more detail, because the components of the "nuisance" can indicate courses of action.

Insects

It has been estimated that ninety percent of the housefly population of our cities breed in open trash barrels and garbage cans or on the food scraps that may spill from them. The larvae can crawl out of the containers within three days of the eggs being laid. Simply ensuring that all trash containers have well-fitting lids would therefore go a long way towards both preventing the laying of the eggs and of the emergence of any larvae from eggs laid at an earlier stage. However, this step is simple in theory and difficult in practice because of public resistance to being "regimented." An easier solution is to employ bags for refuse collection. One of the major attractions to public-health officials of the use of paper or plastic bags is that folding over the top of the bags makes them convenient to carry. Any procedure such as this which breaks into the egg-laying and -hatching cycle is enough to take care of the insect problem without additional care or effort.

Rats and Vermin

Rats are attracted by any available scraps of food, and trash and garbage barrels are a common source. The use of food-waste disposers which grind scraps and discharge into the sewers has greatly reduced the rat population in many areas, although there are unconfirmed reports that rats feeding on food scraps in sewers have been multiplying. The problem of making food scraps unavailable to rats is somewhat more difficult than that of keeping away egg-laying insects. Rats can be clever, strong, and determined. When several commercial repellents, which were supposed to keep rats from biting through paper sacks, were tested by undergraduates at M.I.T. they found some that did in fact deter the test rats [1][1]. Then some local Cambridge rats chewed their way through the wood wall of the laboratory cage, into the treated sack, ate the food and left. Refuse containing food should therefore be stored in strong, closed containers. Paper and plastic bags should be hung in steel-mesh or similar enclosures (Figure 2.1) during use if rats and dogs are able to get into the area, and should be left in the open for collection only for the minimum possible time.

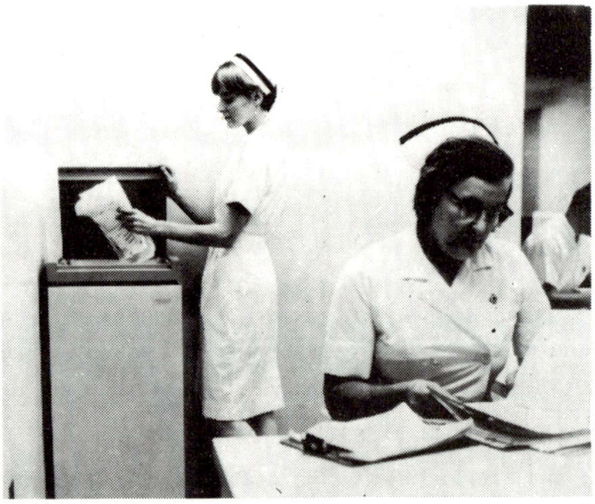

Figure 2.1 Enclosure for paper or plastic refuse bags.

courtesy St. Regis Paper Co.

[1] References for this and later chapters are listed at the end of each chapter.

Fire

Many disastrous fires start in rubbish rooms, trash chutes and trash barrels. Loosely packed dry paper is easily ignited by a smoldering cigarette which may have been cleaned up, for instance, from an ashtray or wastebasket. A fire below an open trash chute is particularly dangerous. Trash chutes should therefore be fitted with fire-stopping "gates" at intervals.

Fire can also start from spontaneous combustion with more densely packed, moister, quantities of trash. These conditions should be avoided, as explained below.

Odor

The sweet but sickly odor of refuse comes from the products of certain forms of bacteriological decomposition. There are very many types of bacteria always ready to go to work in favorable circumstances – that is, favorable to that particular type of bacteria. In dealing with solid wastes, however, we need consider only two general types of bacteria: those that need oxygen to live and multiply and those that flourish where oxygen is not present. The distasteful smell which we have learned to associate with trash comes from the second, anaerobic type of bacterial decomposition. The action of these bacteria is slow and is accompanied by the release of, predominantly, methane gas; it is the form of breakdown that occurs in swamps, in the deeper layers of landfills, and in closed trash bags or containers. The smell can be avoided, and the bacteriological breakdown can be more rapid, if rubbish rooms, trash chutes, trash containers, and barrels are provided with forced-air ventilation. However, it is probably impracticable and, from the fire point of view, undesirable, to blow air through stored trash to maintain aerobic conditions.

Therefore, while correct management should be to ensure that either of two conditions – complete aeration or complete enclosure – is obtained, and never conditions which are almost aerobic or almost anaerobic, aeration will not always be possible. The alternative is to exclude air, because by doing so the danger of fire and infestation by vermin are reduced.

Volume

The most obvious problem of solid waste is that the volume being generated is increasing rapidly. One typical hospital has increased its weekly output of solid wastes from five tons to thirty-five tons in ten years. Most of this increase can be attributed to so-called "disposables" or, more properly, single- or limited-use items. The author of a bulletin of a large plastics raw-materials producer had the honesty to state recently that the problem with many of these disposables is that there is no known way to dispose of them. Such a statement requires careful definition. What was meant was that the plastic materials in particular are so inert that they do not break down when buried in the ground. The toughness and lightness of plastics makes them difficult to shred in any size of pulverizer. It was noticed in one hospital that the vegetation down wind of the incinerator was deteriorating rapidly. It turned out that the chlorine contained in vinyl plastics was forming hydrochloric acid which was attacking almost everything it condensed on. But whether or not limited-use items are noxious in certain circumstances, the large increase in bulk which occurs when an institution switches to paper or plastic products for sheets, pillow cases, uniforms, plates, cups, cutlery and so on adds a severe load to the disposal facilities both on site and beyond, should the institution's wastes be shipped out.

Hazardous Wastes

A hospital or an institution containing laboratories produces solid wastes which may be explosive, corrosive, toxic, or otherwise dangerous. Many an unfortunate janitor has found a cast-off hypodermic needle sticking in his hand after cleaning out a wastebasket. One janitor at M.I.T. was severely burned from a half-full container of strong acid which had been casually tossed into the wastebasket. Diseased animals and specimens of various types form another category that cannot simply be introduced into the general waste stream. Household spray cans are being increasingly used and seem to be becoming more, rather than less, hazardous as flammable or explosive propellants are now freely permitted.

This list of problem areas is not all-inclusive, and most public-works directors and managers of institutions could add their own horror stories to show that particular situations produce particular problems. Dealing with them requires a combination of good techniques with common sense. Complex problems require a combined, or systems, approach. In other words, it is of little use doing three-quarters of a job correctly because the last quarter may be capable of vitiating all the value of the rest of the approach.

In the next section some of the traditional and newer methods of handling solid wastes in homes and institutions will be mentioned.

METHODS OF SOLID-WASTES MANAGEMENT

To overcome all the difficulties stated is a tall order. Some of the approaches which are possible have already been hinted at and will be reviewed along with those that have not so far been mentioned.

While most public attention seems to be concentrated on the final disposition of solid wastes, the principal costs borne by both institutions and by municipal public-works departments are for handling and transportation. Considerable improvements in these areas are possible.

Pneumatic and Hydraulic Duct Systems

The simplest duct system for solid waste is the vertical refuse chute. If designed as part of a building, it is a very efficient way of handling refuse. It can also become foul and can be an efficient way of spreading fire. A refuse chute should be designed with suitable entry chambers and intermediate and final traps to avoid the possibility of it becoming a fire chimney; building and fire codes should be adhered to in the design. Refuse should be bagged or wrapped for deposit in the feed hoppers. Standards for these hoppers can be such as to exclude articles having dimensions larger than certain maximum measurements, thus greatly simplifying the controls for automatically feeding a compactor or incinerator.

The most difficult problem in the planning of the movement of material within buildings or building complexes is to achieve satisfactory horizontal movement. The traditional method is to rely on gravity chutes or elevators for vertical movement, and, for small quantities of materials, hand carts of some variety for horizontal movement. Solid wastes can also be moved through pipes in a liquid or in an air stream. Systems employing these principles have been available for decades, but only recently have the changing economics of capital and labor costs swung in favor of such capital-intensive arrangements in certain circumstances.

Hydraulic Duct Systems

The Garchey system was first introduced in Paris in 1927 [2]. It consists essentially of substituting for the regular sink outlets and drains a flushing unit and larger pipes so that newspapers, bottles, cans, and so on, can be flushed away into the waste system. These solid materials are conveyed by the water to an underground collection chamber, which can be emptied on a regular basis by a special tanker. The overflow from the chamber goes to the sewer in the normal way.

A more recent development of this principle, but one which limits the types of wastes which can be accepted, is to grind the material to a pulp in the sink unit. In this country the Wascon and the Somat Corporations make systems of this type. The pulped wastes are then taken to a hydroextractor where the nonsoluble wastes are pressed against screens and delivered to a container for collection as a compacted semi-dry pulp (Figure 2.2).

These systems are rather obviously more suitable for kitchen wastes, including milk cartons and plastic cutlery and so on, rather than bulkier wastes such as paper bed sheets and magazines which have to be first reduced in size before they can be inserted into the sink chamber. They are not generally sold for home use. The Environmental Protection Agency has recently contracted for the development of a domestic grinder to shred general solid wastes for discharge to the sewer system.

Figure 2.2 Hydraulic waste-handling system.
courtesy Wascon Systems, Inc.

The pulping unit, at the bottom of which is a motor-driven impeller, is in the center. The inclined tube at the right contains a helical screw to press water from the slurry.

Pneumatic Transport Systems

A pneumatic system which, like the Garchey hydraulic system, does not involve grinding, has been developed in Sweden under the name of Centralsug and is being marketed in this country by Aerojet General. Vertical refuse chutes with suitable airlocks run with gently curved bends to a large-diameter (e.g. 24 in.) horizontal transfer pipe. This leads to a turbo exhauster in a central location, which may be as far as a mile away (Figure 2.3). At this point the solid wastes are removed and can be dealt with by any of the traditional or newly developed methods.

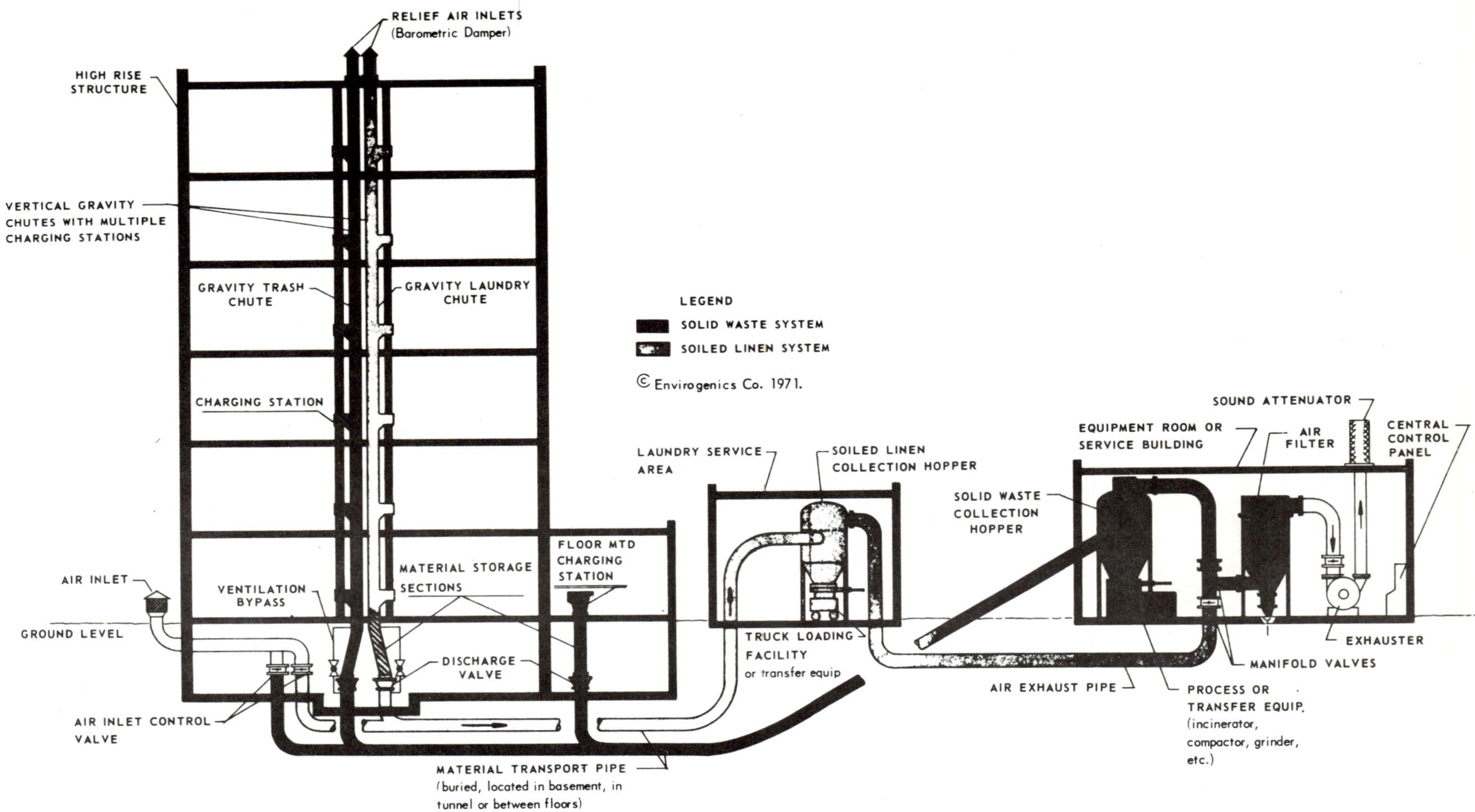

Figure 2.3a AVAC (CENTRALSUG) system diagram.

courtesy Envirogenics Co.

Figure 2.3b AVAC (CENTRALSUG) system test and demonstration unit.

courtesy Envirogenics Co.

This system, like the other duct systems, is an obvious candidate for consideration at the design stage of new buildings, but only in very unusual circumstances would it be worth installing in an old building. More reports have emphasized satisfactory performance. One report was that an automobile lead-acid battery had been dropped, presumably maliciously, into a refuse chute and had been conveyed horizontally to the extraction point without, apparently, causing any damage to the tube. The condition of the battery was not reported.

Iraj Zandi's research into tube transport of solid wastes [3] has led him to conclude that a combination of short-distance pneumatic collection ducts with long-distance hydraulic transportation should maximize advantages.

Containers

The first development in solid-waste management came, one supposes, when the throwing of refuse from one's windows on to the city streets as the principal method of disposal was outlawed. Some form of refuse can was then necessary. However, the word "container" applied to institutional refuse is usually given to something which is larger than a trash barrel and which forms part of a collection system.

Containers come in capacities from, principally, one to thirty cubic yards and in two general categories. Roll-on, roll-off containers, which are normally in the upper end of the size range, are removed bodily by a special-equipped vehicle which may leave an empty container when removing the full container (Figure 2.4). The other general type of container is one which is hoisted in some way and tipped into the rear hopper of a conventional packer collection truck or into the overhead doors of a front-end-loading packer truck (Figure 2.5).

As refuse grows larger in bulk and decreases in density, the costs of using containers with uncompacted trash rise. In the next section, types of compactors and balers are briefly reviewed.

Figure 2.4 Roll-on, roll-off container being hoisted aboard vehicle.

courtesy E-Z Pack Company

Figure 2.5 Front-end loader emptying container.

courtesy Pak Mor Manufacturing Co.

Compactors and Balers

As has been stated, the principal costs of dealing with solid waste are in transportation and handling, and are associated with its volume rather than its mass. The average density can be increased by the simple application of pressure so as to deform or to break boxes, bottles and cans which would otherwise cause large voids (Figure 2.6). Another way of eliminating these

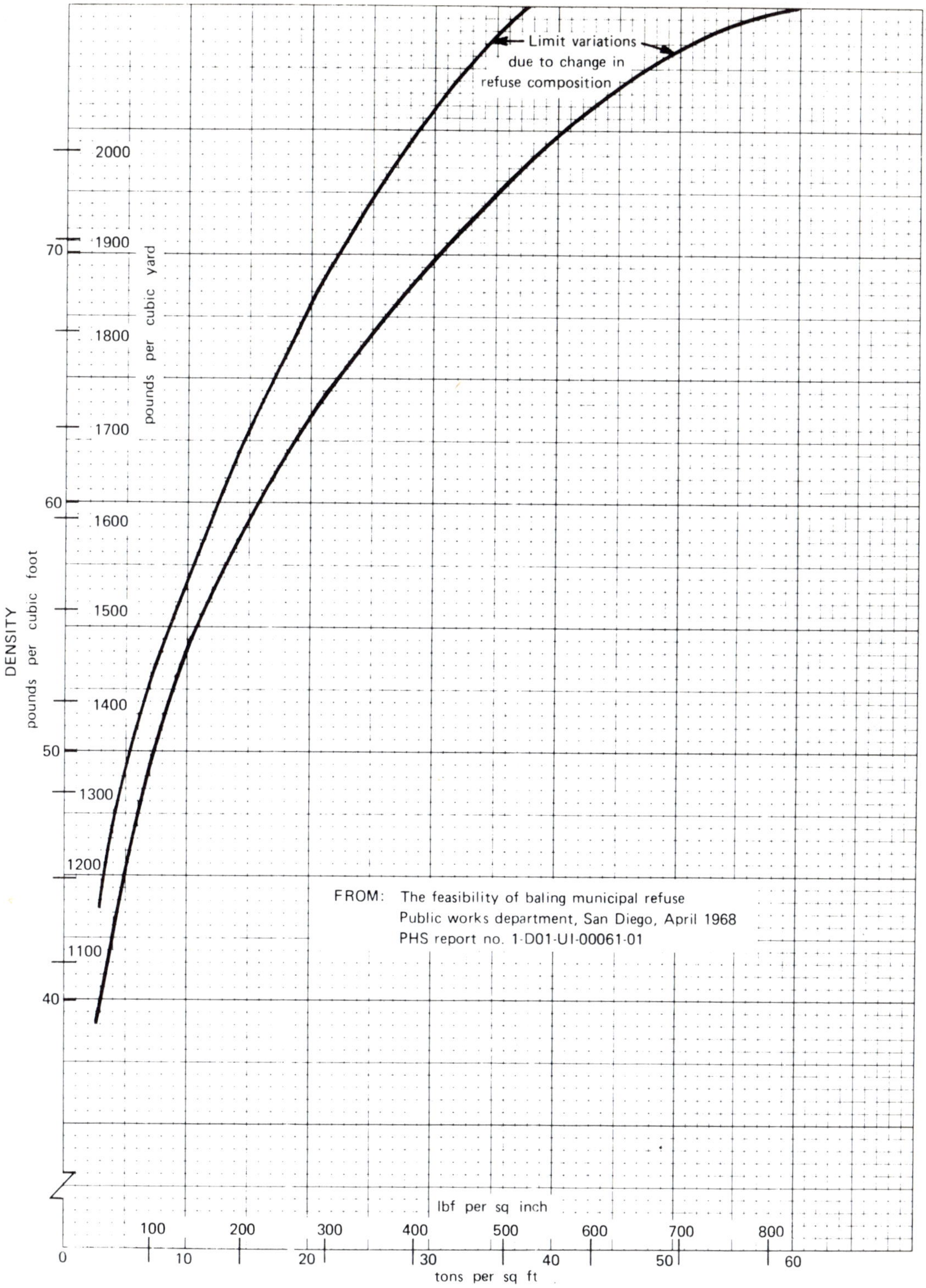

Figure 2.6 Baling pressure versus bale density.

large voids is by reducing the average size of the pieces of refuse by shredding (also known as pulverizing or comminuting), which is discussed in the next section. Small voids remain, and maximum densities are achieved by compaction after shredding.

The great majority of small compactors and balers consist simply of a hopper leading to an open-topped cylinder (Figure 2.7). At intervals a piston or ram advances, trapping some of the refuse and pushing it into a compaction (or compression) chamber. Here there are various design possibilities. In some cases the refuse is pushed into a rigid container, and virtually no paction takes place until the container is almost full. Then the natural springback of dry refuse may limit the amount of additional refuse which can fall into the cylinder. Wetting the refuse reduces springback but adds greatly to the haulage costs, since trucks hauling compacted or shredded refuse are usually weight-limited (axle-load regulations) rather than volume-limited.

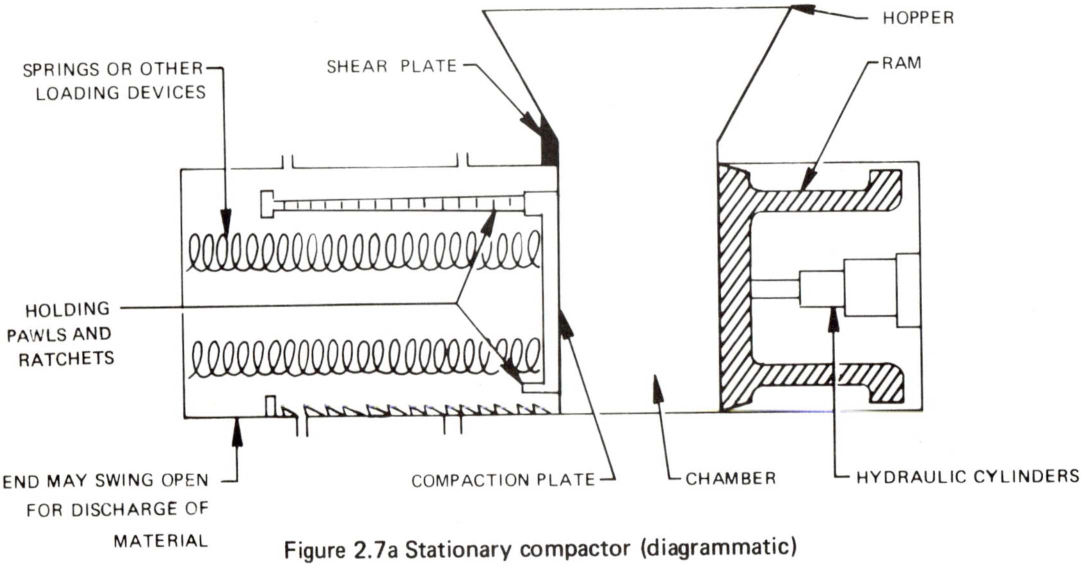

Figure 2.7a Stationary compactor (diagrammatic)

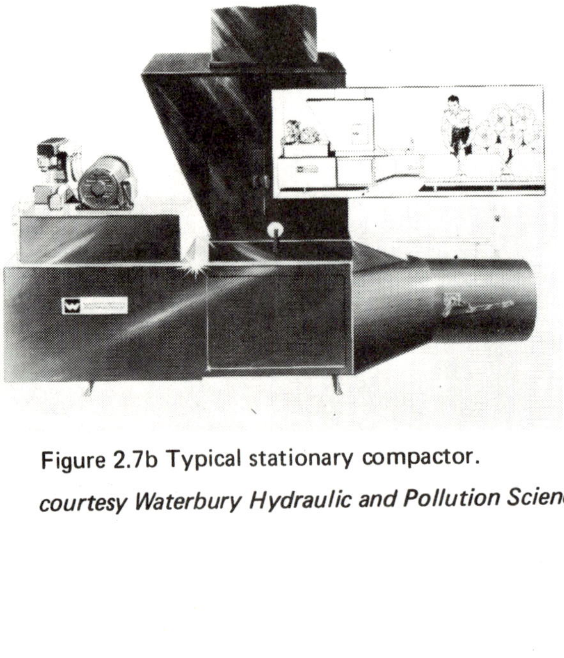

Figure 2.7b Typical stationary compactor.

courtesy Waterbury Hydraulic and Pollution Sciences, Inc.

Accordingly, the ram is often designed to push the refuse against a fixed or moving plate which will ensure that each stroke of the ram compacts the refuse by itself. A greater degree of compaction can thereby be achieved than is the case when pressure is applied only as a last stage. Obviously, a can or bottle surrounded by compliant paper is less likely to be crushed by what is virtually hydrostatic pressure than by a point load applied by a metal plate or bar (Figure 2.8).

Figure 2.8 Advantage of point loading in compaction.

The principal problems of compactors are connected with the feed, while other difficulties are sometimes experienced with the containerization of the compacted refuse. Small compactors, such as the Sears Roebuck Whirlpool undercounter model (Figure 2.9) are hand-filled into containers of plastic-lined paper bags set in a stiff supporting enclosure, and thereby these problems are largely avoided at the cost of considerable hand labor. However, when a compactor is fed from a chute or conveyor, the ram may advance at a time when an article is neither fully in nor is wholly out of the cylinder. If the article is strong and tough, such as a piece of heavy lumber, or a coil of wire, or is fibrous, such as a rug, the ram may advance and not be able to handle the item. Most compactors have a limit switch which, in the case of the Sears Whirlpool machine, for example, reverses the ram direction at a preset load, if this is reached before the end of ram travel.

Two alternative approaches are used by manufacturers to deal with the less-stiff but tough items. One is to leave a wide clearance between the ram and the cylinder so that fibrous materials and wire will simply be pushed aside if they are not captured. The second possibility is to fit the top of the ram and the underside of the forward edge of the cylinder with cutters designed to be able to shear wire and rope.

Instead of pushing the refuse into a rigid container, some compactors compress the trash into individual free-standing bags on a carousel, or into a plastic tube which can later be tied in sections like sausages. At some point the container into which the compactor is discharging becomes full and another container must be substituted. If this procedure is automatic, the machine must be able to deal with anything from a nylon stocking entangled round one object in the compactor and another in the container, to a crow bar which may jam the mechanism. Purchasers of automatic compactors should ask to see tests in which, perhaps, they could bring along some of their own favorite articles of trash. As mentioned earlier, limitations in the size of the feed hoppers to a refuse chute can greatly ease the feed-regulating problem.

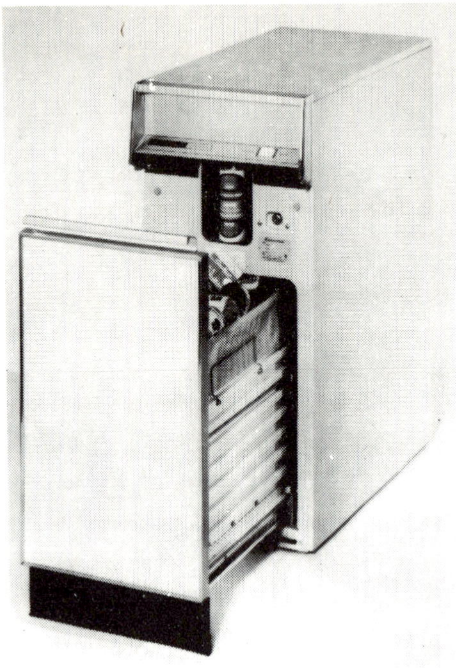

Figure 2.9 Domestic refuse compactor — the "Trashmaster."

courtesy Sears Roebuck & Co.

Shredders

Machines which cut or tear waste into small pieces are available in power ranges from about a horsepower to several-thousand horsepower. The large machines will be discussed in Chapter 6. For on-site waste treatment in offices, apartment buildings and institutions a hundred horsepower forms a reasonable upper limit to the amount of power which can be dissipated in a building which houses non-manufacturing activities.

Power input is not the sole determinant of the suitability of shredding for treating a particular type and flow of waste, but it is important. The reason is as follows. Solid waste is normally so heterogeneous that it is impracticable to maintain sharp cutting edges for size reduction of plastic film of a thousandth of an inch in thickness when the same cutting edge may also periodically encounter a spark plug or a piece of rock. (In contrast, for homogeneous solid waste such as uncontaminated paper, low-powered shredders having sharp rotating knives can be used for proprietary or secret-document disposal – however, the knives can be damaged by the presence of paper clips in the paper.) In order, therefore, to be able to handle a wide range of materials without employing cutting edges, high-velocity impacts are generally used.

The most popular type of solid-waste shredder is the hammer mill, in which rotating arms, fixed or swinging with respect to a hub, impact pieces of trash dropped into their path (Figure 2.10). Small pieces, such as paper and plastic film, are torn because of their inertia and wind resistance. Larger pieces become trapped between the moving hammers and bars (called breaker bars, cutter bars or anvils) and squashed, fragmented or torn in this way.

Since, by this principle, size reduction has to be accomplished at high speed, the power input becomes related to the amount of mechanical work that can be performed in fracturing material. While the design of a shredder has much influence on its capability to handle large or

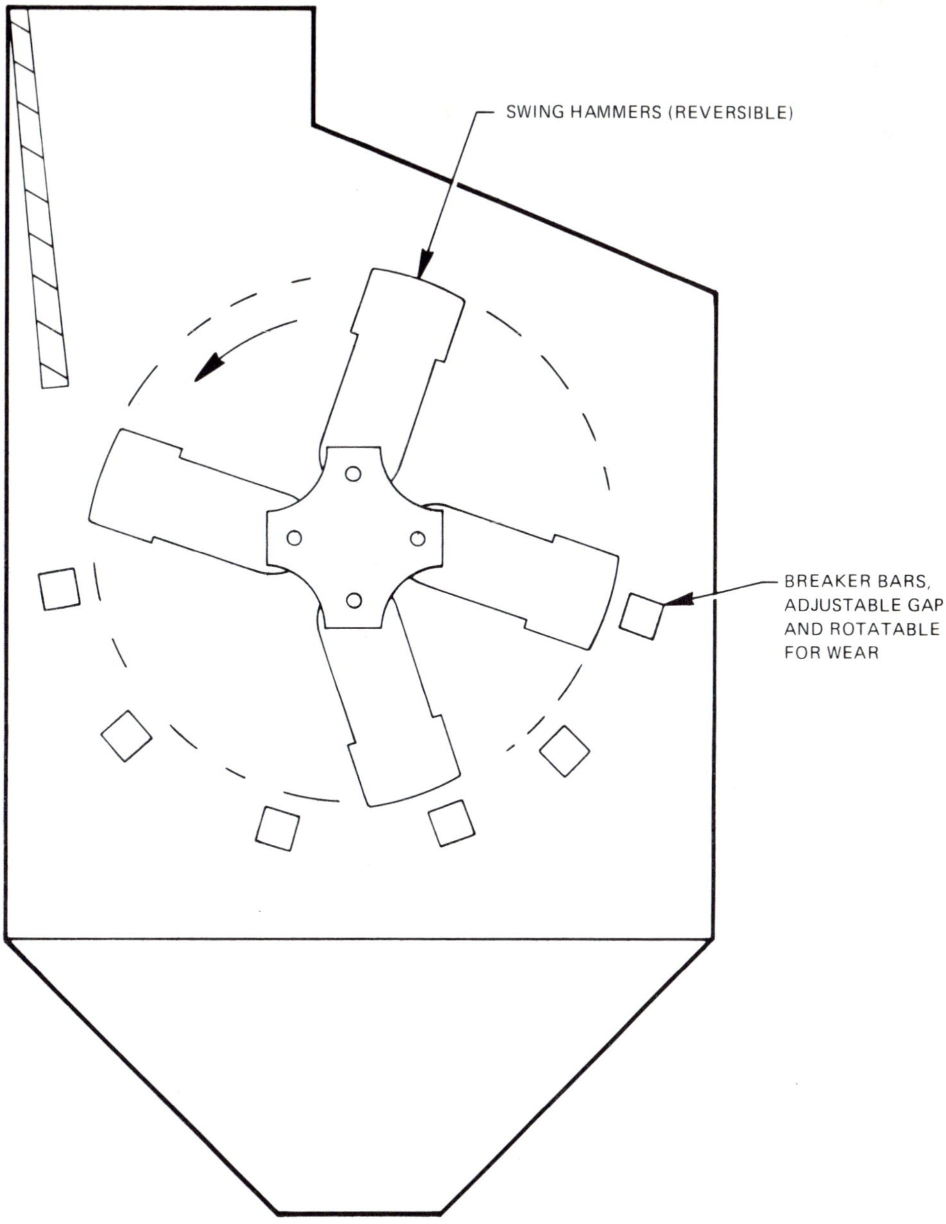

Figure 2.10 Hammer mill (diagrammatic).

tough pieces of refuse (for instance, the rotary inertia of the rotor is important), it is nevertheless a rule of thumb that shredders of below 100 horsepower will become jammed or stalled by perhaps, two-by-fours and 16-gauge steel; one-thousand-horsepower shredders may be able to handle a complete refrigerator or range but will take successfully only the stripped-down hulk of an automobile; and that only with several-thousand-horsepower devices can shredders accept a complete automobile (should this be desirable).

Small shredders are therefore temperamental and are liable to become hung up on coils of wire or rope or rugs, for instance. In an office building one may be able to assume that such articles in the refuse will be rare occurrences, and in this case the rendering of refuse into a material which is easy to store and transport, does not smell, does not attract flies or rats, is resistant to fire, and completely eliminates a prevalent avenue for theft, may combine to give benefits greater than the considerable costs involved. A shredder in this class is the SFM Milpac (Figure 2.11) in which shredding is followed by compression in a hydraulically operated ram. The trash appears as a dense, hard briquette which may be passed without offense around the guests at a supper table.

Figure 2.11 "MILPAK" hammer mill:compactor combination.

courtesy SFM Corp.

Some shredders which rely on the principle of low-speed tearing are under development. If these operate successfully the power limitation described above need not apply: a five-horsepower shredder could conceivably reduce a washing machine.

Incinerators for Buildings

Like shredding, incineration is a process which becomes more reliable and successful as the unit size is increased. While special incinerators may be needed by certain institutions such as laboratories and hospitals for contaminated and noxious materials, a small incinerator is not a good solution for handling general solid wastes. A full treatment of large incinerators is given in Chapter 7.

Salvage and Recycling

It is obvious that within at most a few decades we will have to recycle a large proportion of our solid wastes or reduce either our population or our standard of life drastically. At present our laws and the resulting economics favor the exploitation of raw materials, so that recycling fights an uphill battle in many cases. However, when charges for the removal of solid waste to a landfill or incinerator become high – for instance, ten to twenty dollars per ton – as is the case in many metropolitan areas, it will be well worthwhile trying to determine if secondary-material dealers will at least remove salvageable materials free. Whether or not a price can be charged for these materials depends on two factors: quantity and purity. These topics are discussed more fully in Chapter 8.

REFERENCES

1. Donna Almes, William N. McLeod, and Oakely E. VanSlyke, III, "Paper Cans for Refuse Disposal in Boston," 2.67 report, mechanical engineering department, M.I.T., 1968.
2. Frank Flintoff and Ronald Millard, *Public Cleansing* Maclaren and Sons, London, May 12, 1968.
3. Iraj Zandi and J. A. Hayden, "The Flow Properties of Solid-waste Slurries," *Proceedings* of The International Conference on the Hydraulic Transport of Solids, Cranfield, England, September 1970.
4. "The Feasibility of Baling Municipal Refuse," Public works department, San Diego, April 1968, PHS report No. 1-DO1-U1-00061-01.

CHAPTER 3

URBAN REFUSE COLLECTION AND STREET CLEANING

SUMMARY

A detailed examination of present practices in refuse collection and street cleaning in part of Boston has been made by Lawrence J. Partridge and Joseph J. Harrington of the Harvard School of Public Health, and a summary of their study forms Chapter 4. The study pictures circumstances as they are; the need for change in general is obvious.

The present chapter will summarize some of the particular changes which seem to be desirable; these recommendations are largely those of James A. Freaney.

Chapter 5 will concern one specific area where improvements are underway in some cities across the country: the use of operations-research techniques to arrive at optimum routing strategies for collection and optimum sites for the location of incinerators, transfer stations, and similar centers. This area is treated by David H. Marks.

TYPICAL PRESENT PRACTICE

The variability in current household refuse-collection practice can be judged from Table 3.1, which is taken from [1]. This table represents one page out of about 450 of summarized data. It shows that the area of Boston used as a case study fairly represents conditions across the country, even though in detail the practices followed in Boston form a unique combination. The table shows that the majority of communities which provide household collection service do so once per week, and collect refuse and garbage on that one visit. The predominantly used collection vehicle is a standard commercial chassis having mounted on it a special collection body incorporating a rear hopper of about two-cubic-yards capacity and a compaction system which periodically sweeps the contents of the hopper into the main body and compacts these contents against a moveable ram. A common body size is 20 cubic yards, and when full the payload varies from 3-1/2 to 7 tons of mixed municipal refuse.

The crew size for a 20-cubic-yard packer truck is usually one driver and two or three loaders. Many communities provide one or two "barrel rollers" whose task it is to go ahead of the packer truck and to bring the barrels from the rear of premises to the curbside.

A substantial proportion of communities provide separate collection of garbage, mostly once per week. Separate collections are also made by a substantial proportion of communities for ashes, for combustibles, for noncombustibles and for bulky items.

Costs of urban solid-waste collection as reported in various publications range up to $40 per ton. There are many reasons for this very large range. One is that the accounting systems of many public-works departments and of the types of firms which provide collection services are frequently extremely inexact. Since the figures can seldom be authoritatively challenged, the reported average costs tend to be at the low end of the range when the purpose of the report is to demonstrate the efficiency of the service and at the high end of the range when more taxes or higher fees are being requested.

TABLE 3.1

SAMPLE PRELIMINARY DATA ANALYSIS FROM MUHICH ET AL, [1] 15 AND 16. *HOUSEHOLD REFUSE–COLLECTION FREQUENCY*

	Raw Data	Weighted by 1967 population
Combined Collection		
% Response	97.80	98.39
Sample size	4776	61974068
% 1 per week	68.01	47.33
% 2 per week	23.14	32.14
% Other interval	8.39	20.08
Garbage Collected Separately		
% Response	94.91	97.55
Sample size	1966	41204621
% Not collected	25.08	6.73
% 1 per week	47.05	61.05
% 2 per week	22.02	29.54
% Other interval	5.84	2.66
Rubbish Collected Separately		
% Response	94.98	97.55
Sample size	1975	41075415
% Not collected	26.43	7.83
% 1 per week	47.81	65.61
% 2 per week	14.07	17.29
% Other interval	11.67	9.26
Yard Refuse Collected Separately		
% Response	95.69	97.71
Sample size	2323	44546282
% Not collected	32.92	20.29
% 1 per week	33.10	49.76
% 2 per week	11.60	14.06
% Other interval	22.35	15.87
Ashes Collected Separately		
% Response	94.05	95.70
Sample size	1682	23730330
% Not collected	41.65	23.93
% 1 per week	29.89	44.75
% 2 per week	10.30	17.44
% Other interval	18.07	13.77
Combustibles Collected Separately		
% Response	93.39	94.80
Sample size	1513	19621952
% Not collected	45.01	20.55
% 1 per week	30.64	42.88
% 2 per week	12.88	24.41
% Other interval	11.39	12.04

TABLE 3.1 (Continued)

SAMPLE PRELIMINARY DATA ANALYSIS FROM MUHICH ET AL, [1]
15 AND 16. *HOUSEHOLD REFUSE-COLLECTION FREQUENCY*

Non-Combustibles Collected Separately		
% Response	92.84	95.14
Sample size	1411	21000644
% Not collected	40.07	15.98
% 1 per week	33.12	45.13
% 2 per week	13.05	19.64
% Other interval	13.66	19.14
Bulky Items Collected Separately		
% Response	97.60	98.40
Sample size	4140	63132289
% Not collected	43.10	38.44
% 1 per week	7.15	8.69
% 2 per week	2.52	8.47
% Other interval	47.21	44.39
Total Other Collected Separately		
% Response	100.00	100.00
Sample size	18775	277581417
% Not collected	98.00	98.31
% 1 per week	0.17	0.06
% 2 per week	0.23	0.12
% Other interval	1.58	1.48

Reported collection costs may be for curbside trash-barrel collection alone, or may include street cleaning and collection of garbage, abandoned automobiles, bulky items and so on, all of which generally cost far more per ton than does the collection of general "trash-can" refuse. And lastly, it is much easier, and therefore less expensive, to collect from open suburban communities than from dense city areas where trash barrels must be passed over the hoods of parked cars to the packer truck and where traffic jams slow the progress of vehicles.

POSSIBLE IMPROVEMENTS IN URBAN REFUSE COLLECTION & STREET CLEANING

Combined Collection in Bags

The use of wet-strength paper or plastic bags by residents for combined dry refuse and garbage introduces a number of savings which more than offset the initial cost of the bags.

(1) It eliminates a very expensive separate collection for garbage if such a separate collection has been the practice.

(2) It greatly speeds up collection, or reduces the number of men needed for the same rate of collection, through eliminating the need to pick up and replace a trash can which may contain little more than its own weight of refuse.

(3) The urban fly population, 90 percent of which is estimated to breed on food scraps in uncovered trash barrels, is greatly reduced.

(4) Street litter is greatly reduced because of the general absence of spillage from closed bags.

(5) The noise involved in trash collection is reduced.

Reduction in Crew Size

The use of incentive bonuses and other good management practices has frequently been found to enable a refuse crew to be cut, for instance, from four to two (one driver plus one loader) without loss or reduction in the rate of collection. Going further than this, Ralph Stone & Company [2] has shown that "substantial reductions in the overall costs of providing collection services were possible using. . . .collection systems for curbside collection of residential refuse wherein one man. . .acts as both driver and loader." Side-loading packer vehicles were used for one-man operation.

Quieter Collection Vehicles

With the introduction of paper and plastic bags for curbside collection, the predominant collection noise has become the motor, hydraulic system, and compaction mechanism of the collection truck. Substantial advantages would result if this noise could be greatly reduced. The trucks could then be used on early and late shifts without citizen objections. Because of the reduction in road congestion at these times, the productivity of early and late collections is higher than that of normal daytime collection.

Continuous Loaders

When the rhythm of collection does not have to be broken by the need to wait while the packer-truck hopper is cleared, crews work faster. Some existing vehicles using rotating screws, rotating drums, traveling-chain conveyors, and certain types of blade compaction systems, can be loaded continuously. Some of these arrangements yield an additional advantage in that a degree of shredding, giving improved landfill or burning conditions, occurs.

Integrated Vehicles

The great majority of present packer trucks consist of special bodies mounted on standard chassis. The bodies are therefore high off the ground and leave a great deal of unused space between the wheels. It seems desirable for the packer-truck industry to follow the example of some oil-truck manufacturers in that the container body could serve as the chassis. This would eliminate the frame and would lower the center of gravity, and would probably increase the payload for the same axle load (which is the present principal limit on vehicle size).

Integrated Street-cleaning and Collection Vehicles

If the collection vehicle were also equipped with brushes and vacuum equipment for street cleaning, no additional crews would be needed; the street sweepings would be combined with the collection of solid waste; and parking regulations would need to be enforced for one instead of two occasions. Because of this, community cooperation would be expected to be much better than at present. Collection routes would need to be somewhat shorter because of the reduction in refuse payload, but this cost would be small compared with the substantial savings from the elimination of additional street-cleaning crews and equipment.

Automated Collection Vehicles

The development of a combined sweeping and collection vehicle of the type mentioned, together with the use of bags by householders, would facilitate the introduction of an automatic-loading vehicle (Figure 3.1). This could be operated by one man and would harvest bags or bundles left at the curbside in a similar manner to a wheat harvester [3]. Such a vehicle would be successful only if car parking along one side of a street were eliminated on collection days. To reduce the cost of enforcement, it has been suggested that the drivers of such vehicles could be deputized so that they could issue parking-violation tickets and supervise the removal of violators' vehicles when desirable.

A one-man-operated collection vehicle termed a "barrel snatcher" has been developed by Mark Stragier, public-works director of Scottsdale, Arizona. The city supplies 80-gallon wheeled barrels to householders. These barrels are grasped by a boom under the control of the driver, emptied into the vehicle, and replaced [4].

Figure 3.1 Driver-operated collection vehicle.
courtesy Gulf Oil Corp.

The vehicle is designed to pick up bags and bundles from the curb-side to be deposited in the side-mounted conveyor.

Automatic Loading plus Home Compactors

A further saving in overall collection costs (residents' plus authorities') was forecast in the M.I.T. 1968 summer study by the use by all residents in an area of home compactors so that an automatic loading vehicle would be designed to handle a standard, small, bundle or bag. Since that time, Sears Roebuck and Whirlpool have introduced a compactor, selling at \$200–\$250, which employs a special bag costing 30 to 50 cents (Figure 2.9). An aim for the production of home compactors should be an initial cost of \$75 in the store and the use of standard-size grocery bags, possibly coupled with a light-plastic-bag enclosure or lining or with an aerosol-applied waterproofing.

Bulky-item Container Collection

In a Detroit-area county, a 25–30 cubic-yard roll-on-roll-off container is left once a month in a convenient location in each section of the collection area for the householder to bring along his bulky items – furniture, appliances, and so on. This arrangement practised on a regular basis enormously reduces the cost of special collections of bulky items on demand, or the once-per-year cleanup practised by other communities.

Some communities in the U.S. are using the container system for regular municipal refuse. Householders bring their bags or barrels and empty them into strategically located containers, which may be of the roll-on-roll-off type or may be emptied directly into packer trucks and replaced. Such a scheme for overall collection would be expected to work only in suburban and rural areas where community discipline was high.

REFERENCES

1. A. J. Muhich, A. J. Klee, and P. W. Britton, "Preliminary Data Analysis; 1968 National Survey of Community Solid-waste Practices," Public Health Service Publication No. *1867,* Washington, U.S. Government Printing Office, 1968,
2. Ralph Stone, and Robert P. Stearns, "One-man Collection and the Status Quo" presented at the American Public Works Association 1968 Western Regional Conference and Equipment Show, May, 1968, Las Vegas, Nevada.
3. Igor L. Paul, and David G. Wilson, "Review of Prospects for Automating Urban Solid-waste Collection" an ASME publication, No. *69*-WA/PID-21, November, 1969.
4. Marc G. Stragier, P.E., Public Works Director, Scottsdale, Arizona, "Barrel Snatcher" Proposed for Refuse Collection" *Public Works,* April, 1969.

CHAPTER 4

QUANTITATIVE ESTIMATES FOR SOLID-WASTE MANAGEMENT IN BOSTON

INTRODUCTION

The following discussion is directed toward a quantitative analysis of waste management in densely populated urban areas by means of a case study of the economics and performance of a specific solid-waste-management system.

There are many shortcomings associated with existing methods for handling the flow of solid waste in urban areas. Labor, management and the consumer all contribute to the difficulties involved with providing for waste removal. Ideally, an acceptable waste-management system would ensure that little or no solid waste escapes into the environment. For example, material that eludes the collection cycle can be recovered by complementary maintenance operations, such as lot-and-alley cleaning, street cleaning, and utilization of litter baskets.

Although refuse-collection practices have been described or analyzed in numerous studies [1, 2, 3, 4], very little is known about how solid-waste material escapes the collection cycle and accumulates within the environment. In densely populated urban residential areas substantial quantities of domestic refuse, bulky items, and construction and demolition debris are commonly encountered in uncontrolled conditions. These releases are apparently due to a combination of breakdowns in the collection system, public apathy and disrespect for the established collection procedures, as well as general lack of knowledge and enforcement of the refuse-handling ordinances and contracts concerning both the collection firms and citizens (Figure 4.1).

Little attention has been given to establishing guidelines for allocating manpower and equipment to the complementary maintenance operations that seem to be required at some locations. In order to approach this problem, it is first necessary to estimate the quantities of waste that escape from the prescribed collection system, as well as the productivity of manpower and equipment resources in such operations. A concomitant problem is the need to establish appropriate indices of environmental quality with respect to solid wastes. These would be useful in ranking the performance of alternative programs for solid-waste management. To date, only a preliminary suggestion has been made [5]. This chapter contains some of the data necessary for the development of such indices, but the analysis is preliminary.

This study is primarily concerned with developing a quantitative basis for analyzing such problems. Although this information will be useful elsewhere, the specific details pertain directly to Boston. Where possible, comparisons with other cities are given.

The primary municipal services currently provided to Boston residents for collection of solid waste include refuse collection, street sweeping, and lot-and-alley cleaning. Other operations provided by the Department of Public Works include removal of derelict automobiles, installation and servicing of litter baskets, education and enforcement of policies regarding proper refuse handling within Boston, and operation of an incinerator and landfill.

Figure 4.1 Public apathy leads to litter.

Later sections of this chapter deal successively with the type and amount of wastes generated in Boston, as well as the present allocation of resources and performance of the several solid-waste-management services.

TYPES AND AMOUNTS OF SOLID WASTE GENERATED IN BOSTON

Boston is presently divided into a number of municipal service districts, as shown in Figure 4.2. Some of the distinguishing characteristics of these districts are given in Table 4.1. Since

TABLE 4.1
CHARACTERISTICS OF BOSTON RESIDENTIAL AREAS

District	Population[1]	Area[2] (mi^2)	Street[2] Mileage	Population Density (people/acre)	Median Family Income	Average Family Size
1A	20,000	1.8	22	17	$5,400	2.9
1B	84,000	7.3	95	18	5,900	2.9
2	37,000	5.4	79	10	7,400	3.8
3	98,000	4.1	89	37	6,500	3.7
4	66,000	4.7	65	22	5,700	2.5
5	33,000	2.7	44	18	5,400	2.9
6	66,000	7.1	79	15	9,400	3.3
7	80,000	5.9	89	20	8,900	3.6
8	46,000	5.0	54	14	9,400	3.3
9	30,000	2.7	44	17	5,400	2.9
10	60,000	4.3	95	22	6,200	2.6

[1] How the people see their city, Boston, 1969. Joint Center for Urban Studies, Cambridge, Massachusetts.
[2] Personal conversation with City Engineers Office, Boston, 1970.

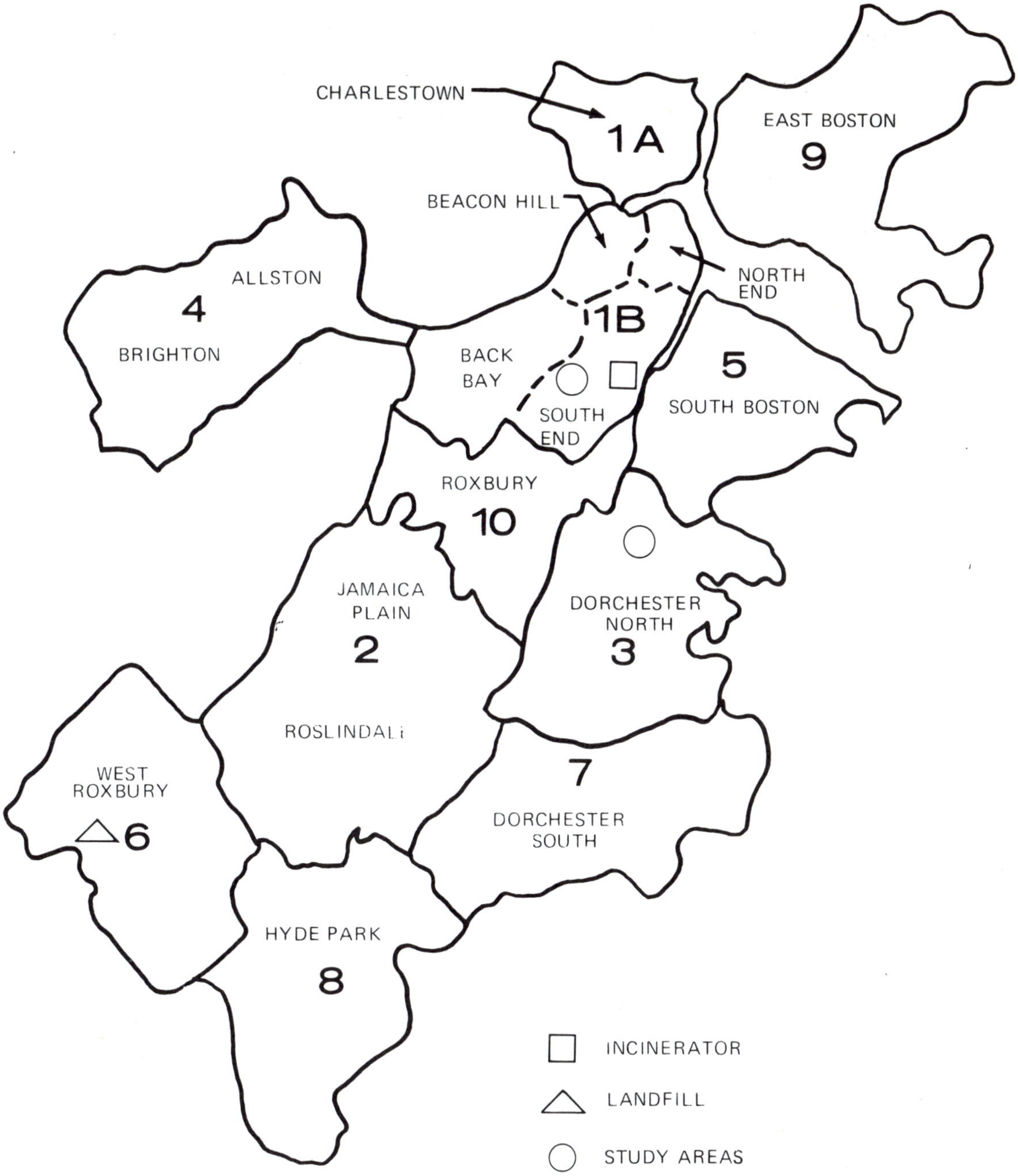

Figure 4.2 Boston solid-waste districts.

the study reported here is directed mainly at the residential environment, the major concern was to estimate quantities of domestic household refuse, discarded bulky items and residue from renovation or demolition of existing structures.

TABLE 4.2
APPROXIMATE COMPOSITION OF DOMESTIC REFUSE AS COLLECTED
(NORTH DORCHESTER)

	Weight	% Total	% Range[1]
Paper	597	41	40-41
Metal	151	10	9-11
Food waste	284	20	16-22
Rags	45	3	1-4
Plastic	25	2	1-3
Glass	312	22	20-22
Construction waste	37	2	1-2
	1451	100	
	(SOUTH END)		
Paper	547	37	26-43
Metal	152	10	9-10
Food waste	312	21	17-21
Rags	48	3	1-5
Plastic	27	2	1-3
Glass	391	26	20-33
Construction waste	19	1	1-2
	1496	100	

[1]Over three sampling cycles.

Domestic Refuse

Gross tonnage figures presented by the Boston D.P.W. indicate that approximately 1,000 tons of refuse are handled daily at its disposal facilities [6]. This amounts to a generation rate of 3.0 pounds per capita daily for Boston. Since this figure includes commercial, domestic, bulky waste, etc, a sampling program was undertaken to verify the generation rates and to determine the composition of the domestic household refuse in two areas judged representative of the densely populated inner city. The areas are sub sections of the South End and North Dorchester. The data gathered during the sampling program will be used in future studies dealing specifically with the two study areas.[1] They are presented here to offer comparisons with the data of others and should not be interpreted as applicable to all of the districts in Boston.

The South End study area contains 50 acres of residential neighborhood, consisting primarily of brick row houses with approximately 4,000 residents. Little open space exists within this district, with the exception of alleyways and scattered vacant lots. The population is approximately 70 percent black, with the remainder consisting of roughly equal numbers of white and Spanish-speaking. The North Dorchester study area (District 3), 195 acres, has a

[1] In addition to the observations reported here, information was gathered by questionnaire during personal interviews. This work was part of a larger project sponsored by the Ford Foundation.

high proportion of vacant land. There two- and three-story multi-family dwellings serve as housing for approximately 7,400 residents, of which 66 percent are black, with the remainder being divided between white and Spanish-speaking.

Random samples of refuse were taken from both areas during September, November and April of 1969-1970. The information in Table 4.2 is based on hand-sorted samples typically consisting of approximately 500 pounds of domestic refuse. The data listed for the Boston survey deviate from those presented by others principally with respect to the glass composition (Table 4.3). The proportions of metals and paper appear to be comparable for all groups. It was also noted that there is an apparent absence of newsprint and magazines in the refuse available for collection in the Boston study areas. Some scavenging of paper and metals was actively practiced in both study regions.[2]

The daily per-capita generation rates for domestic refuse are shown in Table 4.4 and compared with those from other surveys. It is indicative of the present inadequate understanding that no good explanations are available even for large discrepancies.

Bulky Material

The second category of waste material which accumulates within the residential environment is bulky wastes, such as appliances, furniture and wood. The Department of Public Works estimated in 1968 that 75 to 100 tons of such material are collected each week from the entire city [7]. This collection represents the efforts of both municipal and contract haulers. However, this material is not collected uniformly over the year, and intensive clean-up campaigns

TABLE 4.3
REFUSE COMPOSITION IN PERCENTAGES BY LOCATION

Location	Paper	Metal	Food waste	Glass	Construction waste	Rubber textiles plastic
South End	26-43	9-10	17-21	20-33	1-2	2-8
North Dorchester	40-41	9-11	16-22	20-22	1-2	2-7
New Haven, Conn.[2]	33	11	26	16	0.3	13.8
Raleigh, N.C.[3]	39-46	6-8	26-36	11-14	2.0-3.0	3.8-4.2
Rogus[4]	40-46	6-13	5-15	3-9	3-20	3-9
Cincinnati, Ohio[1]	42	9	28	9	2.7	4.0
Memphis, Tenn.[1]	29	7	20	10	1.7	7.8
Johnson City[1]	45	10	26	11	1	4.1
Flint, Mich.[1]	17	14	32	17	1.7	2.8

[1]Niessen, W. R. and Chansky, S. H., The nature of refuse. Proceedings of 1970 National Incinerator Conference, ASME, New York, 1970.

[2]Goody, Marvin, private communication, September 26, 1969, Goody and Clancy, Architects, Boston, Massachusetts.

[3]Partridge, Lawrence, Refuse composition for Raleigh, North Carolina, Masters Thesis, North Carolina State University, 1968.

[4]Rogus, Casimir A., Refuse composition, APWA Reporter, 1963.

[2] It was not, however, possible to make quantitative estimates of the amounts and types of materials scavenged. These operations are not commercial in nature and appear to be conducted on a small scale by neighborhood residents.

TABLE 4.4
DOMESTIC REFUSE-GENERATION RATES IN
POUNDS PER PERSON DAILY FOR VARIOUS AREAS

Location	Mean	Standard deviation	Coefficient of variation
South End	1.70	0.32	0.
North Dorchester	1.40	0.36	0.26
New Haven, Conn.	1.50	0.26	0.17
Raleigh, N.C.	1.55	0.17	0.11
Los Angeles, Calif.[1]	2.59		
Miami, Fla.[1]	1.75		
New York City[1]	2.42		
Philadelphia, Pa.[1]	2.10		
Charlestown	1.52[2]		
Jamaica, Plain	2.60[2]		
Allston, Brighton	1.70[2]		
South Boston	2.28[2]		
West Roxbury	1.73[2]		
Dorchester South	1.89[2]		
Hyde Park	1.82[2]		
East Boston	3.00[2]		
Roxbury	2.40[2]		

[1]Technical and economic study of solid-waste disposal needs and practices, Department of H.E.W., P.H.S., Bureau of Solid-Waste Management, Rockville, Maryland, 1969 Public Health Service Publication 1886.
[2]These values were tabulated from records for truck weights recorded at the city incinerator and truck counts at the landfill. They include both household and commercial refuse but the relative proportions are as yet undetermined.
[3]The mean and standard deviation for the first four entries in the above table were calculated using the following expressions

$$\text{Mean} = \frac{1}{N} \Sigma_i (X_i)$$

$$\text{Standard Deviation} = \sqrt{\frac{\Sigma_i (X_i - \bar{X})^2}{N-1}}$$

X ≡ mean per-capita refuse generation rate for each observation
N ≡ Total number of observations made during sampling program
[3]The calculation for the standard deviation is based upon the assumption that the separate populations comprising the sample have the same mean. See Dixon, W. J. and Massey, F. J. *Introduction to statistical analysis* 2nd Edition McGraw Hill, 1957.

may produce materials at much greater rates. For example, a six-week campaign (summer, 1970), conducted under the auspices of the Model City program, yielded 1,600 tons of bulky refuse [8]. Also, spring clean-up campaigns conducted by the city yield large volumes of bulky material which apparently have been stored by residents in their homes during the winter months. Unfortunately no information is available concerning the composition of this material in Boston. The composition of bulky items collected by the New York City sanitation department is shown in Table 4.5.

TABLE 4.5
PERCENTAGE COMPOSITION OF
BULKY MATERIAL AS COLLECTED[1]

Material	% Total	Density (lbm per cu. yd.)
Trees, stumps, brush	3	230
Furniture, fixtures, appliances	41	130
Lumber, remodeling waste (burnable)	6	355
Cardboard and paper	1	100
Rubbish (burnable)	16	550
Non-burnables	33	175

[1] Bulky and demolition wastes, report made to City of New York, John Baffa, Consulting Engineers, New York, 1968.
NOTE: Composition based upon one-week sampling program for material collected by New York City Department of Sanitation.

Construction and Demolition Debris

The residues from construction and demolition projects are evident in the residential environment primarily as scattered brick, loose timber, and broken plaster and lathing. The amount of residue, Table 4.6, generated by these operations is estimated at 5,000 to 7,000 tons per week, based upon the information presented in Table 4.7 for building and demolition permits in 1970.[3] This type of material creates its major impact upon the environment when indiscriminately dumped on open lots or in alleys. Generally, large-scale demolition sites are reasonably cleaned and graded by the contractors performing the work. However, the quality of the final maintenance program is variable.

Field observations document that loads of plaster and small timbers from renovation of tenements are frequently disposed of in alleys adjacent to the structure. Often this material remains in that location until the city removes it. It is not clear at present whether citizen complaint or surveillance by city inspectors provides the initial impetus for removal of this type of material.

Street Sweepings

The last category of solid waste which concerns us is the most conspicuous and widely dispersed – street residue. Accumulations of waste in the gutters and on the roadways com-

[3] The actual release of wastes does not necessarily take place during the month in which the permit is issued. Records for 1968 and 1969 follow a similar pattern with demolition permits far exceeding those for new construction.

TABLE 4.6
TYPICAL WASTE PRODUCTION (CU. YDS.) FROM NEW CONSTRUCTION AND DEMOLITION PROJECTS[1]

Type of structure	Construction	Demolition
One-family frame or brick	15	160
Two-family frame or brick	20	200
Six-family frame or brick	30	800
One-story 100X200 building	70	4,200
One-to-three-story building	100	4,000

[1]Quad Cities solid-wastes project—an interim report, June 1, 1966—May 31, 1967. Prepared by Quad Cities solid-wastes staff under grant No. DO1-U1-00026. U.S. Dept. of H.E.W. National Center for Urban and Industrial Health, Solid-Wastes Program. Cincinnati, Ohio, 1968.

TABLE 4.7
PERMITS ISSUED FOR CONSTRUCTION AND DEMOLITION IN BOSTON DURING 1970[1]

	JAN	FEB	MAR	AP	MAY	JUNE	JULY	AUG	SEPT	OCT	NOV	DEC
Demolition	27	31	78	47	51	77	48	70	56	50	46	34
Construction	26	18	52	22	21	21	17	10	22	16	19	21

Based upon the above information it is estimated that 5000-7000 tons of debris are removed from Boston during an average week.
[1]Personal communication with City Clerks Office, Boston, Massachusetts, 1970.

prise that portion of solid waste which escapes the collection cycle or is casually discarded by residents and motorists. Also included are leaves, sand and dirt from snow-removal and other operations. The composition of this material is variable and primarily depends upon the characteristics of the area from which it originates. For example, highly industrialized locations generate street residue of quite different composition from that found in residential locations [9]. The typical composition for street residue as collected from various locations is listed in Table 4.8.

The amount of street residue removed from Boston streets annually is approximated in Table 4.9 for 1969. Unfortunately, no reliable data are available regarding the composition of this material. However, observations at district maintenance yards, where the residue is stockpiled prior to removal to the city landfill, indicate that the composition of the residue as collected from Districts 1B and 10 is similar to that represented by heavily populated locations in Table 4.8.

TABLE 4.8
APPROXIMATE ANALYSIS OF STREET SWEEPINGS, PERCENTAGES BY WEIGHT[1]

	Rags	Paper	Dirt	Grass leaves	Glass, stones
Medium residential	2	13	53	23	9
Heavily populated	4	49	24	7	16
Heavy industrial	3	19	54	20	4
Center city (non-residential	1	19	34	37	9

[1] *Manual of street-sweeping practice,* American Public Works Association, Chicago, Illinois, 1966.
Data for Newark, New Jersey, 1955.

TABLE 4.9
AMOUNT OF SWEEPINGS COLLECTED FROM BOSTON ANNUALLY

District	Tons removed	Tons per street-mile	Cost ($/ton)
1A	495[2]	21.8	117
1B	780[2]	8.2	140
	2,920[1]	30.0	208
2	615	7.7	262
3	727	8.2	270
4	508	7.8	360
5	780	17.7	208
6	687	8.7	278
7	1,008	11.3	298
8	405	7.5	410
9	900	13.5	368
10	1,950	20.2	182

Based upon equipment allocations in Table 4.14 and assuming daily collection volumes of:[3]

Mechanical sweepers	9 cu. yds. per day
Hand crews	3 cu. yds. per day
Hokey laborers	0.5 cu. yds. per day
One-man motorized vehicles	1.5 cu. yds. per day

[1] City crews
[2] Private contractor
[3] Assumptions are based upon field observation, communication with the Chief Engineer of Boston Highway Department, and personal conversation with district supervisors and their employees.

Refuse Collection

Primary responsibility for supervising the removal of solid waste in Boston lies with the Sanitary and Highway Divisions of the Public Works Department. These two departments either provide or oversee the services associated with solid-waste management. Services rendered include the administration of refuse collection, lot-and-alley cleaning, street sweeping, servicing of litter baskets, cleaning of the Haymarket area, operation of Boston's two refuse-disposal facilities, and similar services.

Municipal refuse collection is provided on a contract basis, with private entrepreneurs submitting bids to provide this service to any of the 11 designated collection districts (Table 4.10). The per-capita cost for this service varies among districts (Table 4.11). The data for 1969, based upon approximate population and paid bid price, indicate that Roxbury, at $10.30 per person per year, is most expensive, while Charlestown, at $4.55, is the lowest.

It is difficult to evaluate the reason for the variation in collection cost among districts. However, factors [10] which may influence collection costs – collection frequency, population density, and travel distance to disposal facility – are presented in Table 4.12. No trends that might explain the variation of collection costs are immediately apparent. The estimated commercial income for each district is also shown and again offers no basis for the unequal collection costs within the city.[4] The amount of refuse collected per district and the estimated cost for the service in dollars per ton appear in Table 4.13.

TABLE 4.10

1970 EQUIPMENT AND MANPOWER ALLOCATIONS FOR REFUSE COLLECTION

District	Contractor	Collection	Men	Vehicles		
				Packer	Garbage	Bulk
1A	Banco Corp.	Combined	6	2		
1B	James Freaney Co.[1]	Combined	27	6-7		1-2
2	J. Amara & Sons	Separate	13 (3)	3	1 (2)[2]	
3	Jefferies Disposal	Separate	29 (3)	7	2	
4	J.F. Ryan, Inc.	Separate	27 (2)	5	2	
5	Banco Corp.	Combined	12	3-4		
6	M & C Corp.	Separate	20 (5)	4	2 (4)	
7	United Contracting Co.	Separate	21 (7)	7	2 (3)	
8	Howard Disposal, Inc.	Separate	17 (4)	3	2 (3)	
9	Banco Corp.	Combined	15	3		1
10	Dooley Bros., Inc.[1]	Combined	21	6		1

[1] Twice-per-week collection.

[2] Figures in parenthesis indicate possible additions that may be required to provide twice per week garbage collection during the period (6/1 to 8/31) Contract specifications require a minimum of 3 men per collection vehicle. The crew size for the separate collection of garbage may vary from 3 to 5 including driver.

[4] Additional parameters which may explain cost variability including refuse-generation rate and family income are currently being incorporated into a multivariate statistical analysis to evaluate those factors affecting collection costs.

TABLE 4.11
REFUSE-COLLECTION COSTS FOR BOSTON IN 1969

District	Population[1]	Contract bid price[2]	Annual cost per person
1A	20,000	91,100	4.55
1B	84,000	516,100	6.15
2	37,000	326,300	8.81
3	98,000	645,000	6.58
4	66,000	419,900	6.36
5	33,000	164,000	4.96
6	66,000	455,000	6.79
7	80,000	686,000	8.57
8	46,000	288,900	6.26
9	30,000	276,000	9.20
10	60,000	618,000	10.30
Total	620,000	4,486,300	
Average (weighted by population)			7.20

[1] Personal communication with Joint Center for Urban Studies, Cambridge, Massachusetts, 1970.
[2] Contract bid prices as quoted by Sanitary Division, Department of Public Works, Boston, Massachusetts, 1970.

TABLE 4.12
POPULATION DENSITY, HAUL DISTANCE, AND COMMERCIAL INCOME FOR BOSTON COLLECTION DISTRICTS

District	Annual cost[3] per person	Pop. density (people/acre)	Haul distance (one way) (miles)	Estimated[2] commercial income
1A	4.55	17	3.0[1]	75
1B[4]	6.15	18	1.5[1]	9026
2	8.81	10	2.0	484
3	6.58	37	3.5	276
4	6.36	22	4.2	134
5	4.96	18	1.0[1]	827
6	6.79	15	1.0	66
7	8.57	20	3.8	169
8	6.26	14	2.7	71
9	9.20	17	2.5[1]	214
10[4]	10.30	22	2.0	3252

[1] Districts hauling refuse to incinerator.
[2] Untitled report from Sanitary Division of Boston Public Works Department, July 24, 1970. Personal communication with one contractor indicates that these figures may in some instances be low by a factor of ten.
[3] Based upon total contract payments to contractors and estimated population served within each collection district.
[4] Twice-per-week collection.

TABLE 4.13
TYPICAL REFUSE-COLLECTION PERFORMANCE UNDER CITY CONTRACT IN 1969

District	Tons collected[1]	Contract value[2] (dollars per ton)	Tons collected service mile
1A	5,538	16.20	244
1B	27,800 + 34,800[3]	18.47	665
2	17,680	18.43	220
3	28,080	22.90	316
4	20,540	20.39	312
5	13,800 + 3,120[3]	12.16	372
6	21,060	21.60	266
7	27,560	24.90	310
8	15,340	18.83	280
9	16,640	16.58	375
10	33,700 + 11,400[3]	18.50	472

[1] Includes household and commercial refuse. Projected values based upon weight slips from incinerator and truck counts at landfill.
[2] Calculated from 1969 bid contract price and estimated quantity of household refuse collected.
[3] Quantity of commercial refuse collected from respective districts with trucks used solely for this purpose. The volume of commercial business is such that in some districts separate vehicles service only this type of establishment. Where commercial locations are less prevalent the household and commercial refuse are collected with the same vehicle. The value of the commercial contracts is variable among districts (Table 4.12) and probably influences the bid price for any given collection district. Current bid specifications subsidize the collection of commercial refuse from a district by the firm holding the contract for that district. This subsidy is in the form of free disposal for all commercial refuse collected by a firm from within its own district. Competitors are free to compete for the commercial business but must pay the normal disposal fee at city facilities. This subsidy is such that it essentially places monopolistic power in the hands of the firm holding a city contract.

Additional refuse is being collected in Boston by private collection firms not holding a City contract. Approximately 105 collection vehicles are licensed by the Public Works Department to operate in Boston. The amount of refuse collected in these vehicles has not been quantified, and it may not originate entirely from Boston.

Street Sweeping

The street-sweeping maintenance program is primarily conducted by the Highway Division, with Charlestown, North End and the Haymarket areas being contracted to a private firm since 1967. This trial operation was proposed to compare the efficiency of private contractors with that of city forces operating under similar circumstances.

Street sweeping is programmed to occur on the day immediately following refuse collection. A modified form of alternate-side-of-street parking is utilized to provide sweeping the even-number side of the street from 8:00-12:00 A.M. and the odd-number side from 1:00-4:00 P.M. Parking regulations ostensibly to facilitate street sweeping have been permanently posted in Boston Proper, Back Bay, Beacon Hill and Roxbury. Additional areas in the city are scheduled to be posted in the future.

The distribution of equipment and personnel among the 10 districts and their associated annual costs are shown in Table 4.14. Approximate labor, capital, and annual operating costs are listed, together with an estimate of the relative cost, at each district highway yard for providing street maintenance within its district, in Table 4.15.

TABLE 4.14
1970 STREET-CLEANING ALLOCATIONS AT CITY YARDS[1]

	District										
	1B	2	3	4	5	6	7	8	9	10	Total
Supervisors	1	1	1	1	1	1	1	1	1	1	10
Foremen	0	0	0	0	0	0	0	0	0	0	0
Inspectors	11	3	4	3	3	3	4	4	3	6	44
MEO[2]	7	4	5	4	4	4	8	6	5	4	55
HMEO[3]	11	3	5	4	2	4	5	3	4	6	49
Hand crews		2	2	3	2	3	6	2	4	8	33
Patch crews	3	1	1	1	1	1	1	1	1	1	12
Hokey lab.	22	5	5	4	4	4	6	3	6	10	69
FEL[4]	5	1	1	1	1	1	1	1	1	2	15
Sweepers	6	2	3	1	3	3	4	1	2	6	31
Trucks	7	2	4	4	5	4	7	4	6	8	51
Cushmans	4	1	1	0	0	1	2	0	0	3	11

[1] Personal communication with Highway Department, City of Boston, 1970
[2] Machine equipment operator
[3] Heavy-machine equipment operator
[4] Front-end loader

Salary schedule and equipment costs

Supervisor	$10,450	HMEO	$7,384	Dump trucks	$36.00/wk
Inspector	$ 8,216	Laborers	$6,136	Front-end loaders	$74.00/wk
MEO	$ 6,708	Mechanical sweepers	$156.50/wk	One-man haul vehicle	$10.00/wk

Lot-and-alley Cleaning

Lot- and alley-cleaning projects are also the responsibility of the Sanitary Division. Although these two organizations service the entire city, their operation is effectively confined to a few selected locations within Boston. These maintenance sectors are routinely programmed to operate on a cyclic basis, but the volume of requests for their service from Boston's little city halls (local offices for citizen complaints), Office of Public Services and community groups is such that it precludes most attempts to initiate the scheduled service. The approximate allocations of men and equipment for each of these operations are listed in Table 4.16.

The alley-cleaning crew operates almost exclusively in the South End of Boston. Its responsibility primarily lies in the 400-acre tract bounded by Massachusetts Avenue, Harrison Avenue, Clarendon Street and the Penn Central Railroad tracks. Monday-through-Friday refuse collection is provided at approximately 70 locations to remove domestic refuse that private contractors are not required to collect, from the mouths of alleys, where it is deposited by area residents (Figure 4.3). Daily accumulations approach 200 to 300 pounds at some locations, with paper supermarket bags serving as the most popular form of container with cardboard boxes and plastic sacks also utilized. The number and type of container observed at several locations are listed in Table 4.17.

Figure 4.3 Household and bulky refuse in Boston's south end.

TABLE 4.15
APPROXIMATE 1969 STREET-CLEANING EXPENDITURES[1]

District	Labor	Capital[2]	Total	Street mileage	Cost per mile	Annual cost per person
1A	N.A.	N.A.	$ 54,132[3]	22	$2,380	$2.70
1B	44,813	65,987	132,400	95	5,400	7.80
	N.A.	N.A.	514,123[3]			
2	188,304	18,624	206,928	79	2,600	5.55
3	224,232	31,264	255,498	89	2,860	2.60
4	220,404	16,592	236,998	65	3,600	3.60
5	174,840	28,976	203,816	44	4,600	6.25
6	214,268	27,637	241,905	79	3,040	3.60
7	342,188	29,016	281,204	89	4,300	4.75
8	197,815	17,122	214,036	54	3,900	4.65
9	257,792	25,592	283,384	44	6,300	9.40
10	407,216	55,048	462,264	95	4,845	7.80

[1] Based upon data presented in Table 4.14. Allocations of labor and equipment are essentially unchanged from 1969.
[2] Capital costs include the cost of vehicles depreciated at 8% over a five-year period with a resale value of 30 percent of the original purchase price. Maintenance and repair costs are also included in this figure.
[3] Supplemental work performed by private contractor. Separate labor and capital are not available.

TABLE 4.16
APPROXIMATE ALLOCATIONS OF LABOR AND EQUIPMENT
FOR LOT-AND-ALLEY MAINTENANCE OPERATIONS

Labor		
	1	Foreman
	2	Inspectors
	4	Machine-equipment operators (M.E.O.)
	2	Heavy-machine-equipment operators (H.M.E.O.)
	20	Laborers
Equipment		
	4	Trucks
	1	Front end loader (F.E.L.)

SALARY SCHEDULE AND EQUIPMENT COSTS

1 Foreman @ $8,950	=	$ 8,950
2 Inspectors @ $8,260	=	$ 16,520
4 M.E.O. @ $6,708	=	$ 26,832
2 H.M.E.O. @ $7,384	=	$ 14,768
20 Laborers @ $6,136	=	$122,720
		$189,790
4 Trucks @ $1,782/yr	=	$ 7,128
1 F.E.L. @ $3,848/yr	=	$ 3,848
		$ 10,976

Personal communication with Sanitary Division of Public Works Department City of Boston, 1970.

Several commercial firms operating in the South End also benefit from this service by depositing their daily accumulations of trash at the locations serviced by the city. In addition to the daily collection of refuse from alleys, these crews are responsible for maintaining the nearly eight miles of alleys in the South End. Table 4.18 presents the amount of material collected by this service organization in 1969.

The lot-cleaning organization is somewhat more flexible in relation to the areas which receive its designated service. They operate to remove litter and bulky material from vacant lots throughout the city (See Figure 4.4). Table 18 lists data documenting the amounts of material removed by this organization from each of the city districts during 1969.

Although this service is provided throughout Boston, the service is not uniformly distributed (Table 4.19). Approximately 50 percent of all refuse in this category originates in Roxbury and the South End.

Figures 4.5, 4.6 and 4.7 illustrate the proportion of funding provided for removal of solid-waste materials within Boston during 1968. These figures include public expenditures only and do not represent the cost to commercial or industrial firms for removal of their waste material. Contractual arrangements between commercial firms and refuse collectors are not strictly regulated by the city.

Accumulation of litter prior to cleaning

Following cleaning by city

Litter beginning to accumulate following cleaning

Figure 4.4 Vacant-lot maintenance within North Dorchester in 1970.

TABLE 4.17
TYPE AND NUMBER OF REFUSE CONTAINERS DEPOSITED AT SELECTED ALLEY LOCATIONS DURING ONE-WEEK PERIOD

Location	Paper bag[1]	CONTAINER TYPE			
		Cardboard box	Plastic sack	Metal[2] container	Weekly total
1	37	3	7	0	47
2	53	8	6	3	69
3	68	12	16	0	96
4	45	6	21	28	100
5	59	16	22	22	119
6	32	0	18	23	73
Total	294	45	90	75	504
% Total	58.6	8.6	17.9	14.9	
Weight, lbm[3]	1180	495	1170	1280	4125
% Total wt	29	12	28	31	

[1] Bags received at supermarket for groceries.
[2] Generally 32-gallon galvanized metal or plastic.
[3] Total weight of material by category of container.
The average refuse weights were calculated for refuse containers in the South End:

Container	Weight (pounds)
32-gallon metal or plastic	17
Paper supermarket bags	4
Plastic sacks	14
Cardboard boxes	11

The data for 1968 are presented as the most complete and current compilation available, and the relative proportions for each category are essentially unchanged. However, an example of the manner in which costs for providing refuse collections are rising is shown in Table 4.20, where bid contract values for 1968 and 1969 are listed.

Tables 4.21, 4.22 and 4.23 recapitulate the data presented previously.

RESULTS OF FIELD MONITORING OF MUNICIPAL SERVICES

Refuse Collection

There are currently in the literature three approaches to analyzing refuse-collection systems. The first of these [12] is a method whereby the temporal relationship between the various unit operations comprising the collection process are collapsed and average values for each component operation are utilized to evaluate the entire process. There is no attempt to optimize the system; rather, this simple manipulation calculates the time required to collect refuse given the average values for the component operations.

TABLE 4.18
REFUSE REMOVED FROM ALLEY LOCATIONS DURING 1969[1]

Month	Domestic refuse		Bulk material	
	Tons	% Total	Tons	% Total
January	80	11.2	110	10.2
February	43	6.0	37	3.3
March	85	11.8	51	4.6
April	61	8.4	60	5.5
May	55	7.7	122	11.3
June	41	5.7	75	7.0
July	45	6.3	79	7.3
August	53	7.4	96	8.7
September	56	7.8	109	10.1
October	55	7.6	117	10.6
November	62	8.6	111	10.2
December	82	11.5	113	10.2
TOTAL	718	100.0	1,080	100.0

[1] Based upon information received from the Sanitary Division of the City of Boston, 1970. Figures for 1970 are approximately the same for domestic refuse with substantial reductions in the amount of bulk collected. The reduction of 32 percent is indicative of positive results from this operation which began in 1968. The large amounts of bulk prevalent prior to 1968 have now been removed and the reduced amounts collected (1970) may better reflect the incidence of this material in the South End. The amount of domestic refuse collected is more constant owing to its relatively uniform generation rate and to the low tolerance level by residents for its accumulation within the environment.

The second approach [2, 3, 4] employs the technique of simulation analysis to retain the temporal relationship between the separate unit operations. Assumed statistical distributions for the system input parameters are sampled to provide data for the simulation. The analysis typically proceeds by varying system inputs and examining the resulting output to establish the most efficient combination of variables (truck size, crew size, location of transfer station, etc.). The results generally include measures of system performance under varying conditions of system parameters. These may be used for selecting among alternative designs.

The third method [1] employs multiple-regression techniques for evaluating the cost of providing refuse collection as a function of several system parameters. The statistically insignificant variables are deleted from the defining expression, yielding an equation for the collection cost as a function of one or more "explaining" parameters.

The following analysis adopts the first approach [12], although additional studies of the second and third type above are underway.

Comparisons between crews or districts are not directly possible. Rather, the economics of collecting refuse in a typical densely populated urban setting are based upon a three-week time-and-motion survey of a refuse-collection crew conducted during February and March, 1970. Days in which inclement weather – snow or rain – may have influenced either refuse characteristics or crew performance are presented separately.

TABLE 4.19
NUMBER OF LOTS CLEANED AND TONNAGE OF REFUSE REMOVED FROM VARIOUS DISTRICTS OF BOSTON IN 1969[1]

DISTRICT	LOTS CLEANED		REFUSE REMOVED	
	Number	% Total	Tonnage	% Total
1A	357	8	88	4
1B	407	9	444	19
2	152	3	93	4
3	823	20	391	17
4	122	3	79	3
5	658	16	185	8
6	55	1	39	2
7	191	4	142	6
8	118	3	89	4
9	253	6	102	3
10	1,141	27	715	30
TOTAL	4,277	100	2,367	100

[1] Based upon personal communication with Sanitary Division of City of Boston, 1970.

Figures for 1970 exhibit overall reductions of 20 percent in the average amount of refuse collected per lot. As with the alley cleaning operation, this probably reflects positive results from this program. Initial accumulations of material have been removed and the city is able to service more lots with less material being removed per location.

Refuse collection in the South End is provided twice weekly, with pick-ups being made primarily from alley locations by 20-cubic-yard rear-loading collection vehicles. Collection crews consist of a driver (chauffeur) and two loaders (shakers). The chauffeur's job consists almost exclusively of driving the vehicle, with the exception of very heavy stops, at which he will temporarily become a loader.

Shakers who have worked together for a period of time function rather well as a team, and gradually develop work patterns which tend to speed collection. For example, at stops where there are several cans located 8-10 feet from the truck (Figure 4.8) crew member (I) takes a can from location 1 to 2. His partner (II) empties the can and places it at location 3. Meanwhile the next can has been moved to location 2, and the empty at 3 has been returned to 4 by crew member II. Often two cans will be carried from the locations 1 to 2, and 3 to 4. However, it is unusual to see one man lift two barrels into the truck simultaneously.

The proximity of collection stops is such that the shakers tend not to call the truck to a stop at a location where only one can is out. They would rather pick up the container and carry it a few steps to the next pickup location, where the contents are deposited in the vehicle. The container is thereby displaced somewhat from its original location.

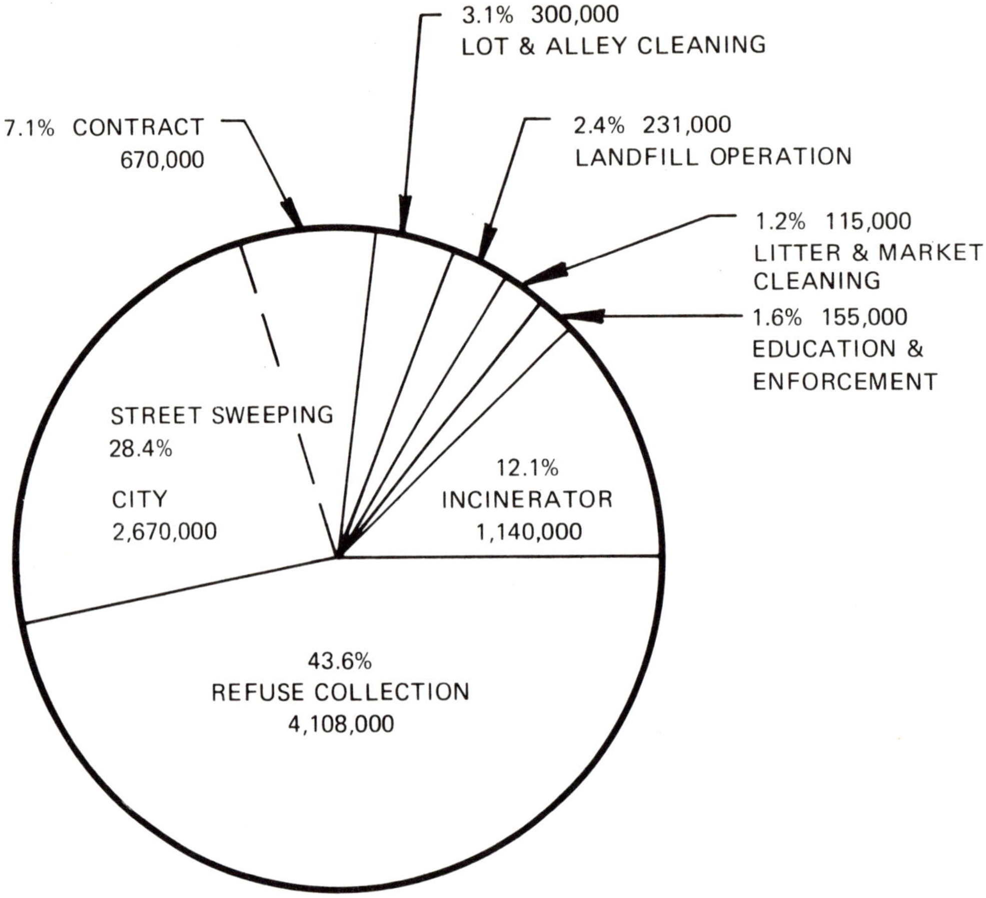

Figure 4.5 Typical total expenditures for solid-waste management in Boston (includes administration costs).

The collection crew is rather knowledgeable concerning the contract specifications regarding quantity, composition and designated pickup location for the refuse they collect. They may not, in fact, collect wastes in non-standard refuse containers.[5] Some residents are unaware of the type and quality of service they are designated to receive under the city contract. For example, many residents are unaware that bulky materials are collected in special vehicles which often must be requested for removal of certain materials. It may be that the information system itself is deficient and that more accurate understanding of the contractors' and residents' responsibilities would improve the situation.

The data in Table 4.24 were recorded while a particular collection crew was observed for three weeks. During this period both the conventional rear-loading packer and a more recently developed continuous-loading truck were employed for collection.

[5] The sanitary code for Boston specifies the required type of refuse container as being of metal or other durable material, not larger than 20 inches in diameter, nor more than 28 inches in height, with a tight-fitting cover. Unfortunately this regulation is widely abused and poorly enforced, as is typical in most large cities.

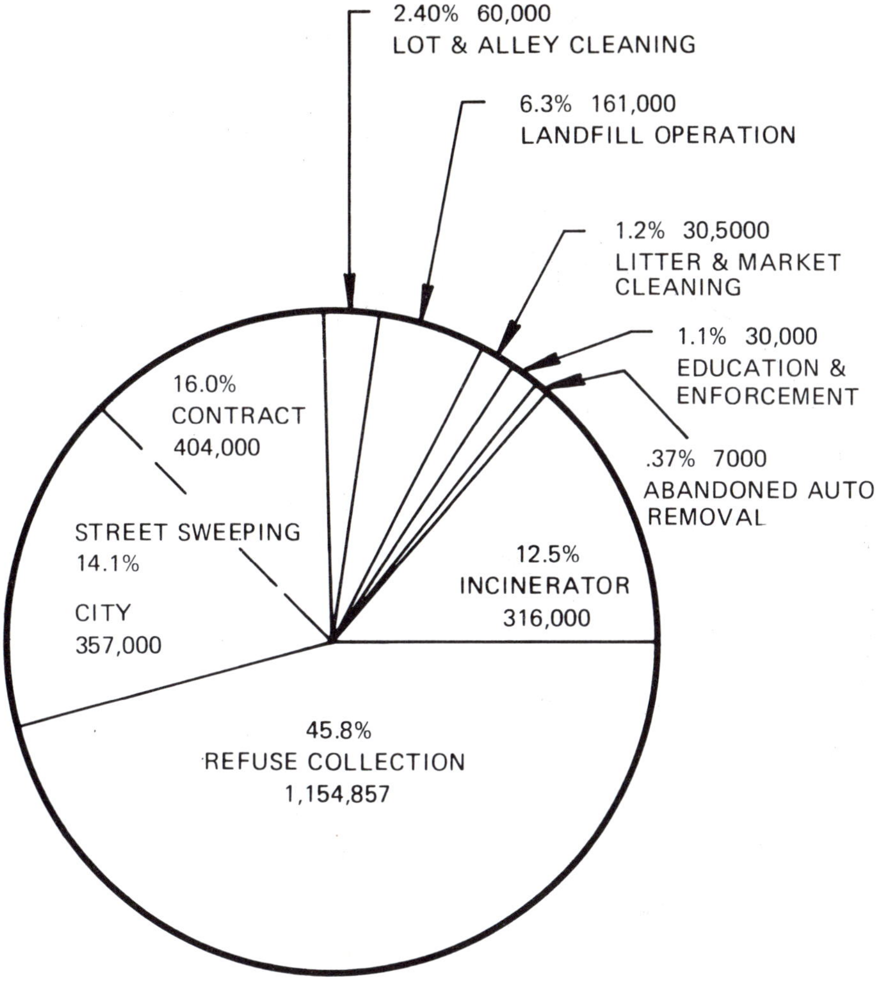

Figure 4.6 Typical capital expenditures for solid-waste management in Boston.

The basic expression for man-minutes per ton can be formulated as follows:

(1) Total man-minutes per trip, $f = b + ade + aT' + aT'' + aT'''$

- $a \equiv$ men per vehicle
- $b \equiv$ pickup time in man-minutes per trip
- $c \equiv$ average tons of refuse per trip
- $d \equiv$ mileage to and from disposal site
- $e \equiv$ average haul speed in minutes per mile
- $f \equiv$ total man-minutes per trip
- $x \equiv$ total man-minutes per ton, that is, f/c
- $T' \equiv$ disposal time, minutes
- $T'' \equiv$ off-route time (coffee break, breakdowns, personal time)
- $T''' \equiv$ travel time to first stop of day

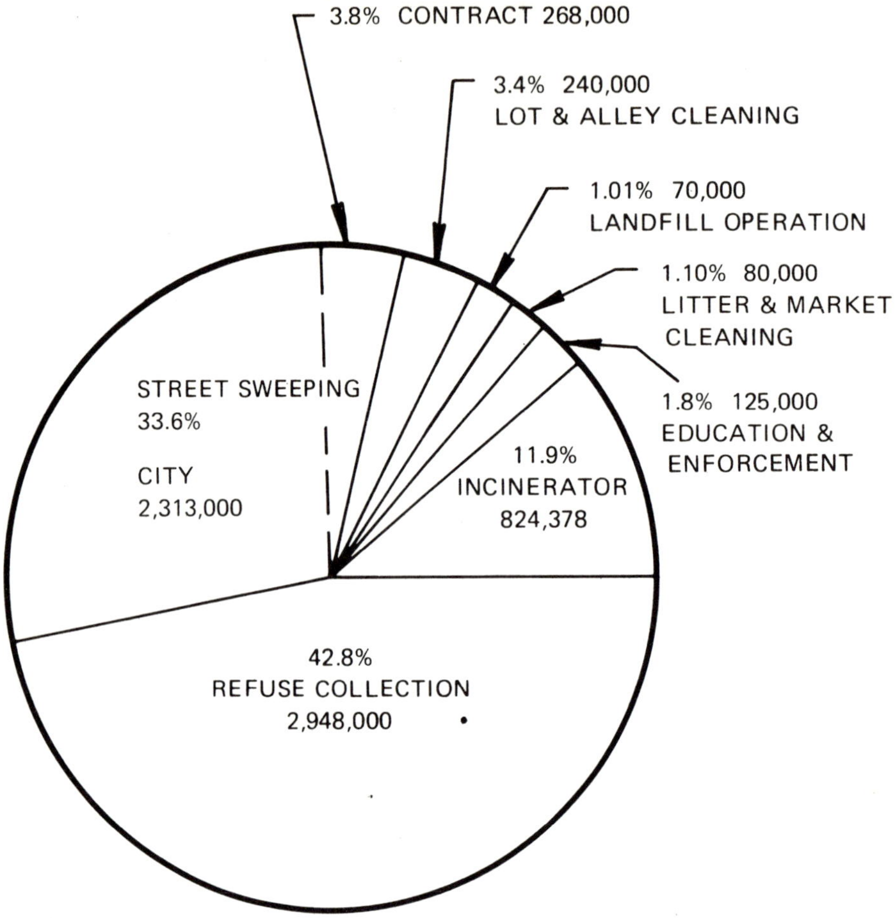

Figure 4.7 Typical labor expenditures for solid-waste management in Boston.

Information gathered for Boston during the time-and-motion survey for twice-per-week collection from alley locations with three-man crews is as follows:

b = 275 man-minutes per trip
a = three men per vehicle
de = 9 minutes to disposal site
T' = 3 minutes disposal time
T'' = 32 minutes off-route time
T''' = 9 minutes travel time to first stop
c = 3.30 tons average refuse per trip
f = (275) + 3(27) + 3(3.0) + 3(32) + 3(9) total man-minutes per trip
x = 130 man-minutes per ton

$$\text{Labor cost per ton} = 130\ \frac{\text{M-M}}{\text{ton}} \times \$.085/\text{MM} = \$11.00/\text{ton}.$$

TABLE 4.20
REFUSE-COLLECTION COSTS
IN BOSTON IN 1968 and 1969

District	1968 Cost	1969 Cost	% Increase
1A	$ 84,479	$ 91,100	6.6
1B	402,655	516,600	28.2
2	263,194	326,300	23.9
3	566,054	645,000	13.9
4	380,870	419,900	10.4
5	149,213	164,000	9.9
6	403,654	455,500	12.8
7	591,518	686,000	15.9
8	229,777	288,800	20.9
9	259,424	276,000	6.4
10	515,437	618,000	19.9
TOTAL	3,849,525	4,486,300	11.5

1970 contracts were performed under the 1969 schedule with 75 percent of the $0.25 per hour increase in labor cost being absorbed by the city. The contract negotiated by the sanitation men in April of 1971 granted them a 58 percent wage increase over three years.

TABLE 4.21
PER-PERSON QUANTITIES OF SOLID WASTE
REMOVED DAILY FROM BOSTON (pounds)

District	Refuse collection	Street sweeping	Lot residue	Total material collected
1A	1.52	0.17	0.02	1.71
1B	1.85 + 2.3[2]	0.24[1]	0.03	5.05
2	2.60	0.12	0.01	2.73
3	1.57	0.05	0.02	1.64
4	1.70	0.05	0.01	1.76
5	2.28 + .52[2]	0.17	0.03	3.00
6	1.73	0.07	0.01	1.81
7	1.89	0.09	0.01	1.98
8	1.82	0.07	0.01	1.90
9	3.00	0.21	0.02	3.23
10	2.40 + 1.0[2]	0.23	0.06	3.69

[1] Includes work of private contractor.
[2] Commercial refuse collected by special commercial vehicles.

TABLE 4.22
TONS OF SOLID WASTE REMOVED ANNUALLY FROM BOSTON

District	Refuse collected	Street[2] sweeping	Lot[3] residue	Total
1A	5,538	495	88	6,121
1B	27,800 + 34,800[4]	3,700	444	66,744
2	17,680	615	93	18,388
3	28,080	727	391	29,198
4	20,540	508	79	21,127
5	13,800 + 3,120[4]	780	185	17,865
6	21,060	687	39	21,786
7	27,560	1,008	142	28,710
8	15,340	405	89	15,834
9	16,640	900	102	17,642
10	33,700 + 11,400[4]	1,150	715	46,965
	277,290	10,975	2,367	290,400

Residue collected by the alley-cleaning operation totaled 1,790 tons which originated entirely from the South end (District 1B)

[1] Refuse collected under city contract.
[2] Residue removed from streets during approximately 40-week period.
[3] Material collected during 1969 and includes household refuse and bulky material.
[4] Commercial refuse collected separately from domestic refuse. Verified by examination of weight slips from incinerator.

An additional calculation employing the data of Table 4.24 gives information on cost per refuse-collection stop which may be a more meaningful descriptor of system performance than cost per ton.

The average collection time per trip was 432 man-minutes, during which 82 locations were serviced. The hourly rate of pay for the crew was $3.87. Under the daily-route method, however, whereby these men are free to leave following completion of their assigned task, they were paid at a rate of $5.10 per hour of actual work. This hourly rate yields a cost per stop of $0.45 for labor.

Additional costs to the contractor that are related to labor include workman's-compensation insurance and contribution to the workman's health-and-welfare fund.

Equipment costs can be distributed in terms of capital, maintenance and operating costs. Additionally, garaging and other overhead costs must be included. These costs are undoubtedly variable, depending upon the type of equipment employed, its condition, and type of maintenance program conducted by the firm.

There is no disposal fee for refuse collected under city contract. The agreement with the city also provides that any refuse collected by a contractor from his district may be disposed of without charge at any city-operated disposal site. The normal (commercial) fee for disposal is $5.00 per ton at the incinerator and $1.00 per cubic yard at the landfill, with only combustible material being accepted at the incinerator.

TABLE 4.23
ANNUAL PER-PERSON COST FOR SOLID-WASTE MANAGEMENT WITHIN BOSTON[1]

District	Population	Refuse-collection cost	Street-sweeping cost	Lot-cleaning cost	Total ($/person)	Median family income
1A	20,000	4.55	2.70	.37	7.62	5,500
1B	84,000	6.15	7.80	.45	14.40	5,900
2	37,000	8.81	5.55	.21	14.57	7,400
3	98,000	6.58	2.60	.34	9.52	6,500
4	66,000	6.36	3.60	.10	10.06	5,700
5	33,000	4.96	6.25	.47	11.68	5,400
6	66,000	6.79	3.60	.05	10.44	9,400
7	80,000	8.57	4.75	.15	13.47	8,900
8	46,000	6.26	4.65	.15	11.06	9,400
9	30,000	9.20	9.40	.30	18.90	5,400
10	60,000	10.30	7.80	1.00	19.10	6,200
[2] Average		7.2	5.12	.32	12.70	

[1] Exclusive of disposal costs and those costs for removing derelict automobiles, cleaning litter baskets, education and enforcement, etc. Expenses for these services are approximately $3.15 per person annually when averaged over the entire city.
[2] Averages are weighted by population per district.

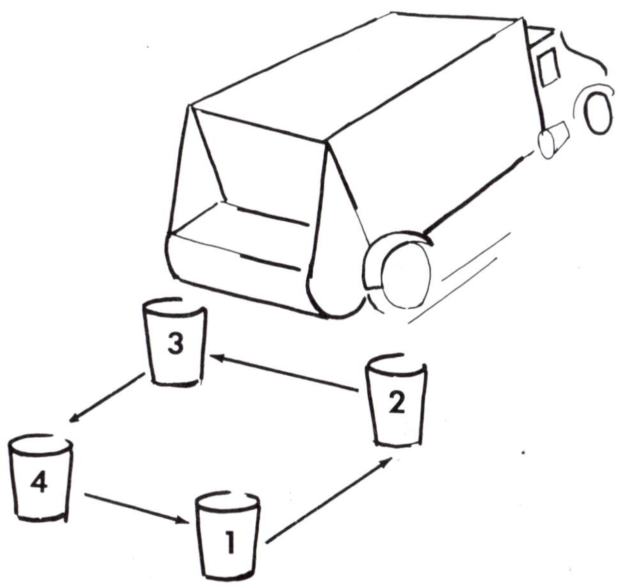

Figure 4.8 Cooperation in a team of collectors.

TABLE 4.24
PARAMETERS FOR DOMESTIC REFUSE COLLECTION
WITHIN THE SOUTH END DURING
FEBRUARY AND MARCH OF 1970

	Mean[1]	Standard Deviation
Weight per trip	7219	1299
Stops per trip	82	28
Collection time per stop[2]	1.7	0.44
Travel time per stop	1.6	0.50
Pounds of refuse collected per stop	82	29
Service density (stops per mile)	130	35
Travel time from yard to route	27	6.75
Travel time to incinerator	27	5.10
Travel time from incinerator to route	27.3	4.10
Time to dump refuse	9.0	3.60
Off-route time	45	12
Break time	45	

[1] Mean and standard deviations were calculated using the following formulae
[2] All times are recorded in man-minutes

$$\text{Mean} = \frac{\Sigma_i X_i}{N} \qquad \bar{X} \equiv \text{Mean}$$

$$N \equiv \text{Number of observations}$$

$$\text{Standard deviation} = \sqrt{\frac{\Sigma_i (X_i - \bar{X})^2}{(N-1)}} \qquad X_i \equiv \text{Individual observation}$$

The individual components which comprise the refuse-collection costs per stop to the contractor are listed below:[6]

Labor 0.45	0.45
Workman's compensation	0.08
Capital cost of vehicle	0.06
Registration, insurance excise tax	0.03
Fuel, oil, maintenance	0.05
Taxes (federal, social security, state unemployment)	0.02
	$0.69

[6] Present contract does not include paid sick days. Vacation time is not included in the analysis.

Assuming that the contractor's overhead and profit are 4-10 percent and 10 percent of the contract [10, p. 374] value, the total cost to provide the refuse collection is approximately $0.76-$0.82 per stop.

The average number of persons per pickup location in the South End is 14, based upon data gathered in a questionnaire survey conducted concurrently with the sampling program. Applying this figure to the cost results indicates that the per-person cost to the contractor for providing this service is $5.60-$6.00 per person per year.

On the national level [13], 3 pounds per person per day of domestic refuse is generated, with an associated collection cost of $13.40 per ton. Comparable figures for Boston in general and the South End in particular are $19.80 per ton, 1.90 pounds per person per day, and $18.50 per ton and 1.85 pounds per person per day, respectively. Therefore, on an average basis, refuse generation in Boston is somewhat lower than the national average, with collection costs being higher. Comparative data for other cities using three-man crews collecting from curb or alley locations are presented in Table 4.25.

Refuse-collection costs in Boston are generally higher than those for the municipalities listed in Table 4.25. This is difficult to explain because Boston is served by private contractors (one of the few major cities to have such an arrangement), who, being profit-oriented, are expected to be more efficient than their municipal counterparts. The reduction in costs when employing private street-sweeping firms has been demonstrated in Table 4.9. However, the same results do not occur for refuse collection.

Information presented by Ralph Stone and Company [11] showed that for cities with more than 500,000 people the average cost per ton collected was $24, with a median cost of $12.78. The average cost is misleading because it is unduly weighted by two large cities with extraordinary costs. Boston's average cost of $19.90 is higher than the median but lower than the average.

It is possible to make an attempt at estimating the effect that inclement weather has upon both the collection and haul times, and therefore the cost per stop. Data collected during a day that was preceded and/or interrupted by rain indicated that the cost per stop was $0.53, an increase of 18 percent over the average for the study period.

The average collection and travel time per stop during this period increased by approximately 24 and 19 percent, respectively. Precipitation had a marked effect on the weight of refuse collected per stop. The 193 stops on this particular day had an average weight of 110 pounds, an increase of 32 percent over the study average, which may be attributed to rain.[7]

Rain can also have an appreciable influence upon disposal. It has also been noted that following periods of abnormal precipitation substantial quantities of refuse accumulate at the incinerator, providing several days' backlog because of reduced burning rates with wet material. The quality of the residue during these periods is also poorer than that normally encountered.

Street Sweeping

The maintenance operation of street sweeping is programmed to function throughout the year. However, the extreme weather conditions during December, January and February result

[7] A standard 32-gallon metal or plastic container left uncovered may collect 8 pounds of water during a 1/2-inch rainfall.

TABLE 4.25
REFUSE-COLLECTION COSTS FOR CITIES WITH POPULATIONS GREATER THAN 400,000 USING THREE-MAN CREWS FOR CURB OR ALLEY COLLECTION[1]

Population	Tons collected	$/Ton	$/Capita – Year
525,000	202,400	8.90	3.42
920,000	340,000	8.80	3.25
700,000	242,800	7.60	2.62
8,100,000	2,843,000	28.20	9.80
870,000	460,000	9.70	5.17
418,000	166,500	11.60	5.60
(Boston) 620,000	320,000	14.00	7.20

[1] National solid-waste survey. Bureau of Solid-Waste Management, Environmental Protection Agency. Cincinnati, Ohio, 1970.

in limited operations during these three months. Employees who are typically assigned to street-sweeping details during favorable weather conditions are utilized for snow-removal and other maintenance work during the winter [14].

Most of Boston is programmed for once-per-week sweeping of streets. Boston Proper, with daily service, and Back Bay, Beacon Hill and Roxbury, with twice-per-week cleaning, deviate from this pattern. Based upon this schedule, city forces are responsible for sweeping approximately 1,200 street-miles per week.

City workers have 31 mechanical sweepers available, of which two-thirds, or 20, may be operating on a given day and of which 15 are laid up for three months during the winter. Typical mechanical sweepers operating in Boston can maintain 8 to 10 miles of street per day. Under average conditions this operation results in the collection of 9 cubic yards of debris. Labor, capital, and operating cost during this period are $28.40, $20.00 and $13.00, respectively, for a total expenditure of $61.40, or approximately $6.40 per mile swept. Depreciation of sweepers is about $100 per week, based upon a first cost of $17,500, useful life of three years, an assumed interest rate of 8 percent, and salvage value of 30 percent initial cost. Applying this figure for capital depreciation to the inoperable and unused sweepers.[8] results in a cost estimate of approximately $8.65 per mile swept or $346/per mile-year for once-per-week service.

Mechanical sweepers can maintain 75 percent of the allocated street miles, with 300 miles per week being the responsibility of hand-sweeping crews. This average of 60 miles per day is distributed among 33 hand crews, 69 laborers with pushcart and broom, and 10 one-man motorized vehicles. Each hand crew consists of three to four laborers plus a truck and driver. Considering an average of 175 laborers and 40 trucks are employed daily, this operation results in direct costs of almost $5,500/per day.[9] The cost per service-mile is $91, or $3,640 annually,

---

[8] Over the course of one year these unemployed units account for approximately 47 percent of available sweeper capacity.

[9] The current daily rates (1970) for laborers and truck driver are $23.60 and $25.80, respectively. Truck costs are assumed to be $7.20 per day.

per mile, based upon a 40-week period. Additional methods employed for maintaining the streets include the use of hokey laborers (one man with pushcart and broom) and the use of one-man motorized Cushman vehicles. The respective productivity for each of these four operations is assessed in Table 4.26.

TABLE 4.26
PRODUCTIVITY OF VARIOUS STREET-CLEANING OPERATIONS FOR BOSTON

Cleaning Method	Average Daily Removal (cu. yd.)	$/Ton
Mechanical sweeper	9	45
Hand crews	3	282
Hokey laborer	0.5	306
Cushman vehicle	1.5	88

The figures shown above are only approximate, but give an order of magnitude for the variable costs associated with maintaining streets free from residue. Other considerations when choosing cleaning methods would include hindrance by parked cars, dead-end-street situations and so on, which serve to complicate street-maintenance programs.

There are additional costs to this system which are not included above but bear mention. These include the cost of posting parking regulations, towing and ticketing of illegally parked vehicles during periods of street cleaning, and maintenance of district highway yards. The Highway Department performs other functions in addition to street sweeping, and it is not clear what portion of each respective budget contributes to overhead at district yards. For this reason, overhead associated with buildings and administration is not included in the cost estimate.

Lot Cleaning

The performance of the lot-cleaning crews was evaluated by observing and photographing their operating procedures in North Dorchester during May and June of 1970, and documenting the size of each lot cleaned, amount of material removed, crew size, type of equipment required to provide the servicing, and time required to complete the assignment.

Lots were classified on the basis of public or private ownership. The lot sizes, amount of material removed and cost for the service are listed in Table 4.27.

Examples of some typical lot characteristics and expenditures involved in removing accumulated waste are listed in Tables 4.27 and 4.28. A stepwise multiple linear-regression approximation for the cost of cleaning lots in North Dorchester during June and July of 1970 is of the following form:

TABLE 4.27
CHARACTERISTICS OF LOTS CLEANED BY
BOSTON PUBLIC WORKS DEPARTMENT IN 1970

	City owned	Privately owned
Mean area, sq. ft.	6,800	6,600
Std. deviation, sq. ft.	3,060	3,800
Mean refuse accumulation	0.40 lbm/sq.ft.	0.36 lbm/sq.ft.
Std. deviation of refuse assimilation	0.35	0.21
Average cost of cleaning	$0.01/sq.ft.	

TABLE 4.28
TYPICAL VALUES FOR LOTS CLEANED AND
THE COST OF PROVIDING THIS SERVICE

Lot area sq. ft.	Material removed (lbm)	Labor cost	Equipment cost	Time spent (hr)
11,000	8,000	$146.80	$29.40	6
15,000	2,000	57.00	21.60	3
3,000	14,000	218.20	88.40	8
3,600	2,000	46.45	8.10	3
31,000	8,000	425.75	36.00	13
22,000	8,400	302.00	95.60	
59,000	7,000	285.60	21.60	8
14,000	6,000	225.00	26.40	4

$$Y \equiv \alpha + \beta_1 X_1 + \beta_2 X_2 + \beta_3 X_3$$

$$Y \equiv 39.92 + 0.028(X_1) + 0.001(X_2) + 0.045(X_3) \qquad (1)$$

where $Y \equiv$ Cost of cleaning in dollars
$\alpha \equiv$ Regression constant
$X_1 \equiv$ Weight of material removed from lot in pounds
$X_2 \equiv$ Area of lot cleaned in ft^2
$X_3 \equiv$ Time since last cleaning in weeks
$39.92 \equiv$ Intercept
$r \equiv 0.836$

The t-test[10] for significance indicates that β_1 and β_2 are significant at the 0.001 level, with β_3 being not significant.

[10] Snedecore, G. W., and Cochran, W. C., *Statistical Methods,* 6th Edition (2nd printing), Iowa State University Press, Ames, Iowa, p. 391, 1968.

Data for the entire city in 1969 show that 4,374 lots were cleaned, with approximately 2,288 tons of debris being removed. The cost for this service is estimated at $46 per lot cleaned and $87 per ton of material removed. These figures appear to be somewhat lower than would be predicted for North Dorchester based upon Equation (1) above.

The gradual degradation of lots to their initial condition was commonly observed. Generally, the accumulation of refuse and other debris on a lot appears to be related to its physical setting, type of structures surrounding it, amount of pedestrian traffic over its surface, and its perceived function in a given neighborhood. No quantitative evaluation of this process in time is yet possible.

Alley Cleaning

The operation of the alley-cleaning crews was observed and evaluated in the South End during June and July, 1970. Selected alley locations were photographed daily prior to and following servicing by the city. The time required to remove the trash from each location was recorded, along with an estimate of the weight of refuse removed.

The estimated average weight, based upon the data presented in Table 4.17 and an enumeration of the type of container at each service location, was 100 pounds with a standard deviation of 37 pounds.

The average time required to clean a location – frequently including the use of brooms and shovels because of the scattered nature of the residue – was 48 man-minutes, with a standard deviation of 16 man-minutes. The estimated mean cost for providing this service and the travel is $4.82.[11]

Periodic cleaning of the entire lengths of alleys is also conducted by this organization utilizing different crews and equipment. The workers providing daily service at the mouths of alleys are generally not the same individuals who clean the entire lengths of the alleys. A typical South End alley is approximately 400 to 500 feet long and could contain domestic refuse, discarded bulky items, as well as illegally dumped debris from renovation projects. The average maintenance crew [12] can remove an estimated 4,000 pounds of material from a single location during an 8-hour period. The cost for providing this service is $257.50 per alley cleaned and $128.75 per ton of residue removed.

It should also be mentioned that this material is widely dispersed, requiring that it be aggregated into a reasonable pile before it can be loaded into the trucks for transport to the disposal area. Also, fences on both sides of the alley restrict the maneuverability of the heavy equipment often required to remove large objects. The city has faced suits for damage or destruction of private property resulting from attempts to move trash away from fences and walls. The impact of such suits has been to restrict the use of heavy equipment where there is a question of whether the material can be more safely removed by hand. The results are that hand labor is being employed where motorized equipment might be more efficient.

Here the question arises whether or not the party on whose property the refuse has been allowed to accumulate should be allowed to file suit for any damages suffered during attempts to remove the trash. However, there is seldom proof that the owner is the person responsible

---

[11] On the basis of a crew of three laborers and one driver in a packer truck. The hourly rates for the laborers and driver are $2.95 and $3.55, respectively. The truck is evaluated at $5.25 per hour.

[12] The crew consists of one inspector, two drivers, one loader operator, four laborers, one supervisor, one front-end loader and two dump trucks. The hourly rates for these categories are $4.30, $3.55, $3.65, $2.95, $5.00, $1.88, and $0.90, respectively.

for the trash; he may be the double victim of having his property first used for miscellaneous dumping and then having it damaged during the removal of trash. This is one area where the present system for handling the waste load is simply not capable of coping with existing conditions and might be examined further to provide some recommendations for future implementation in the area of alley cleaning.

During 1969 this alley-cleaning service was provided on 231 days, during which 710 tons of domestic refuse and 1,080 tons of bulk material were removed from the South End. Based upon the data listed in Table 4.16, the cost to the city for providing this service was approximately $100 per ton of material collected, exclusive of disposal costs.

EXAMINATION OF PRIVATE-CONTRACT STREET CLEANING

The William Banfield Corporation is currently providing street maintenance for both Charlestown and Haymarket-North End on a bid-contract basis. The bids were awarded at a monthly rate of $4,511 and $13,450, respectively. The contracts provide for the collection, removal and disposal of street residue from these areas for a one-year period beginning April 1, 1970.

The contract for the North End-Haymarket area requires one sweeper to work two shifts daily for five days per week. One spare sweeper should also be available. In addition to the two sweepers, one front-end loader, two dump trucks, and four laborers must be supplied to clean the market area on Saturday and Sunday.

There are 22.7 miles of street in Charlestown which are maintained for $54,132 annually. The contract calls for 56 hours of sweeping weekly, with allowance made for maintaining one spare sweeper. Payments for private service are $2,380 per mile swept, $18.65 per hour, or $2.70 per person annually. These figures compare favorably with those estimated costs for each district served by city employees. The average cost per person and per mile for city maintenance is $5.10, with a range of $2.60 to $9.40. The average per-mile cost for city work is $4,200, with a range of $2,600 to $6,300.

However, it should be mentioned that this contract calls for mechanical sweeping, but no mention is made of employing hand labor. A great portion of the cost of maintaining Boston streets is involved with salaries to hand laborers. If only mechanical sweepers were used, the cost of maintaining streets would be lower and perhaps more nearly equivalent to those of the private contractor operating in Charlestown.

SOLID-WASTE DISPOSAL IN BOSTON

The City of Boston currently uses incineration and landfilling in the processing and disposing of household, commercial and bulky wastes collected within the city. On-site incineration, barge incineration, and metal shredding and salvage are also employed to dispose of domestic refuse, construction and demolition debris, and automobile hulks, respectively. Additionally, approximately 8,500 tons of separately collected food waste is being sent to piggeries at a cost of $30,000 per year.

Domestic and commercial refuse collected under city contract is brought either to the South Bay incinerator or to the West Roxbury sanitary landfill on Gardner Street on the banks of the Charles River. Approximately equal amounts are processed at each site, with 195,000 of 350,000 tons processed at the landfill in 1968 [15]. Relative costs for operation of these facilities are at two extremes, with disposal costs being $9.82 and $1.04, respectively, at the incinerator and landfill (1968 figures).[13]

City Incinerator

The city incinerator, built in 1959 at a cost of $5.5 million, has been plagued with problems [16]. Rated at 900 tons per day, it has never processed at that rate, and averages only 425 tons daily. Residue quality is often poor and remains attractive to flies, rodents, and gulls, with some paper and garbage remaining identifiable after processing. This is due, in part, to short furnace-residence times, which average only 20 minutes. Also, furnace temperatures of 1,100 F are 300 to 700 degrees F lower than the optimum for non-slagging incinerators; higher temperatures result in fewer odor and secondary-smoke problems. Following quenching, the residue consisting of non-burnable cans, glass, ash and partially consumed materials, is deposited adjacent to the incinerator prior to trucking to the landfill for final disposal. Originally slated for use as a final cover material, the generally poor quality of the residue restricts its use to intermediate cover, with clean cover used as the top layer.

This facility accepts combustible household and commercial refuse generated from five city collection districts. There are limitations on the size of the material accepted: all bulky items, even if combustible, must be disposed of at the landfill. Material which is collected by the lot-and-alley cleaning crews and which meets specifications is also burned at the incinerator. Hospitals, state and city agencies have disposal privileges at this facility, but other firms or individuals must pay $5.00-per-ton disposal fee[14] on any material collected within Boston. The majority of the material processed is that collected under city contract. Refuse disposed of on a fee basis generally ranges from 10-40 tons daily, which generates an income of $50-$200. Solid waste originating from outside the city is not accepted for disposal.

West Roxbury Landfill

The 98-acre area, site of an old dump, was first used by the City in its present capacity in 1958. The landfill operation has been controversial at times, with complaints being registered against odors, fires and potential contribution to the pollution of the Charles River. However, there have been no fires at the landfill during the past six months.

The operating capacity at the landfill was stated to be 195,000 tons in 1968, but it is uncertain how this value was established because no scales exist at this site. In addition to refuse collected from six of the 11 city districts, all non-combustible material, as well as street sweepings, are deposited at this location. Debris from demolition or construction and automobile hulks are not allowed at this location because of its limited future life. Additionally, state and city agencies, as well as hospitals and schools, have free dumping privileges at the landfill. Refuse collected from outside Boston is not allowed. However, firms not holding city

---

[13] Proximity to disposal site was the criterion employed by the PWD when designating disposal sites for material collected under city contract. However, material collected from within Boston but not under city contract may be disposed of at either location. Under the current fee system this policy may in fact influence the choice of site utilization, depending upon the density of the collected material. For example, a 25-yard vehicle carrying refuse at 300#/yard would pay a fee of $25.00 at the landfill and $18.75 at the incinerator. The same vehicle with a load of 1000#/yard would be charged $62.50 at the incinerator, with no change in the fee at the landfill.

[14] This fee is approximately one-half of the disposal cost incurred by the city in operating the incinerator. Such a policy may be based upon the rational of maintaining low disposal fees to encourage private use of this facility, thus reducing illegal dumping in Boston.

contracts but collecting commercially in Boston are allowed to dump their material for a disposal fee of $1.00 per cubic yard. City records show that approximately 90 non-contract vehicles deposit 400-500 cu. yds of refuse daily at the landfill. Fees for this service average $400-450 daily. The majority of this income is received from firms disposing of non-combustible commercial refuse. City residents, who have free disposal privileges, do not use the site to any great extent.

An estimated five years of life remains, even though it is filled with compacted refuse from 20 to 60 feet deep over its entire area. There is apparently no plan for constructive use of the site after the filling operation is abandoned. Unfortunately, the lack of foresight in this matter is common in this region. Two exceptions to this are the former disposal areas at Columbia Point and the Fort Point Channel, which are currently being used as building sites.

The Public Works Department is actively pursuing opportunities to purchase land adjacent to the present landfill location. There is some concern about the amount of pollution which the present site is contributing to the Charles River, which directly abuts the fill. No quantitative study of the pollution potential has been made. But it would be primarily caused by percolation and runoff from the fill area entering the river. From other studies, it may be anticipated that these flows would be high in suspended solids, and would have a high biochemical oxygen demand, iron, chlorides, etc. [17]. In addition, the use of adjacent land sites for fill locations would require a variance from the state on its recently passed legislation prohibiting the filling of either tidal marsh or inland wetlands [18]. In the event that a variance is not granted by the state, the city may be forced to consider other sites somewhat removed from the metropolitan area. Even with complete conversion to incineration for processing of domestic refuse, the city will require acquisition of additional landfill sites for disposal of non-combustible material and residue.[16]

Alternative Disposal Possibilities

Additional sinks outside of city jurisdiction which receive refuse from Boston include the DeMatteo dump in Saugus, which receives domestic, commercial, and industrial refuse; the Charlestown and Malden barge operations of Fournier Marine Corporation, and the Aeroc Company, both of which companies burn construction and demolition debris at sea and return the residue to shore; landfills in Billerica and Lynn (Massachusetts) and Pelham (New Hampshire)[15] which accept demolition debris; waste-paper firms in Massachusetts and New Hampshire, which process corrugated cardboard; and the Proler Plant in Everett, which shreds automobile hulks and sells the scrap principally to Japan. Future possible processing facilities for solid waste include ship incineration of domestic refuse [19] and ocean dumping of compressed solid waste. The compression baling plant of Reclamation Systems, Inc., in Cambridge may also be used.

There is very little information regarding the volumes of materials processed or the economies involved in any of these sink operations. The Saugus dump receives approximately 1,200 tons per day of refuse from 17 surrounding communities at a disposal fee of $1.65 per ton.[17] The Fournier barge operation employs four barges, which transport approximately 5,000 tons

[15]The one-way haul distances from Boston are 20, 8 and 30 miles, respectively.

[16]This facility having a processing capacity of 2000 tons per 16-hours operation was severely damaged by fire and is being rebuilt with a scheduled reopening for March 1972. However, the major problem facing the successful operation of this facility is locating a suitable disposal site for the baled refuse. To date (March 1972) no disposal site has been secured.

[17] [17]This location had been operated under emergency provision since 1967. However, the owner of this site has been granted a permit from the local board of health to operate this site in compliance with public-health regulations.

per week of demolition debris to sea, where it is allowed to burn uncontrolled on the anchored barges. After burning, the residue is allowed to cool before being returned to shore. The entire cycle takes several days. Charges are estimated at $30 to $50 per 30-cubic-yard load, but are believed to vary. Aeroc utilizes one barge at approximately 1,000 tons per week and is assumed to operate under economic situations similar to Fournier. Public hearings were held before the Department of Public Health licensed these operations, and there is currently some clamor regarding the environmental pollution being caused.[18] The fill sites in Lynn and Pelham accept demolition and construction debris at $40 and $15 per 30-cubic-yard load, respectively. No information is available concerning either the daily capacity or life expectancy of either site.

Clean fill material may be deposited free of charge at Logan Airport, but the nature of the buildings being demolished in Boston is such that a great degree of control must be exerted to ensure combustible and non-combustible materials remain separated. Again, no information is available regarding the amount of residue being disposed of at this location.

The major contractors currently engaged in demolition work within Boston express concern over the lack of suitable disposal sites for their debris. Boston, with the assistance of federal funding, is undertaking urban renewal on a large scale. In 1970, 615 structures were demolished and since 1960 approximately 500-700 buildings per year have been torn down.

The cost for disposal of this material is borne indirectly by the city because disposal fees are reflected in the bid demolition price. The gradual removal of any or all of the present disposal sites will no doubt be reflected in increased bid prices for demolition work resulting from increased haul distances to any new sites. No data are available for the amount of demolition being undertaken in the private sector.

The Proler Plant in Everett has capacity for processing of approximately 1,400 vehicles per day [20]. Estimated costs for shredding are $10 per vehicle, with the resulting scrap being priced at $21 per ton [21]. This plant seemingly has sufficient capacity to absorb all retired vehicles from New England for some time into the future [22]. No figures are available for the number of vehicles retired annually in Boston, but city forces collected 400 to 500 in 1969 [23]. This figure probably amounts to no more than one percent of the total abandoned annually in the city.

On-site Treatment

On-site processing of solid waste occurs primarily in flue-fed incinerators, which in their current state could probably be covered by a blanket statement of being poor waste-processing devices. Some 75 tons per day of domestic refuse are processed in the 444 incinerators at 44 public-housing projects [24]. Additionally, approximately 120 tons per day are processed in 1,128 other incinerators throughout the city. The distribution and capacity of these incinerators is shown in Table 4.29.

The Boston Air-Pollution-Control Commission is currently developing a ban against installation or utilization of on-site incinerators [26]. The potential results of such legislation is unknown; but such a ban, as demonstrated by similar action in New York [27], would increase the disposal problems. Elimination of on-site incinerators in Boston would thrust an additional 180 to 200 tons per day of refuse into the waste-collection and disposal system.

The Commonwealth of Massachusetts has recently issued a document regarding allowable emissions from incinerators [28], but the effects of this legislation, on either promoting

[18] Aeroc has recently been prohibited from further operation because of a poor performance record.

TABLE 4.29
ON-SITE INCINERATORS

Location	Total No.	Capacity (tons per day)
Commercial	140	18
Apartment	808	80
Schools	180	18
Public housing	444	75
Total	1,572	191

incinerator obsolescence or encouraging upgrading of existing facilities to continue present processing rates, are not predictable.

Food-waste grinders (garbage disposals) are not widely used in Boston. There is no legislation against their installation, but the Public Works Department does prohibit the installation of any apparatus which could contribute to sewer blockage.[19] The Metropolitan District Commission, which operates the waste-water-treatment facility for the Boston area, has recently drafted guidelines concerning allowable releases of material into sewers feeding into its treatment facilities [29]. It may be that food-waste grinders will be prohibited in at least some sections of the city.

Several commercial establishments are employing baling operations for processing waste cardboard. This material is apparently being sold for $11 per 30-cubic-yard container of compressed material [30]. This is probably the most attractive alternative for those firms faced with upgrading of incinerators in order to comply with air-pollution-control laws.

Planning

Only 50 percent of the waste generated in Boston is disposed of in city facilities. Thus it is imperative that the city engage in planning to provide for some urgently needed disposal facilities. The closing of any or all of the disposal areas outside the city would require that either the city or some private organization engage in emergency plans for new refuse-disposal sites. The Saugus dump has been operated under emergency legislation since 1967, and many individuals would like to see it closed immediately. Also, incineration at sea, as well as disposal in neighboring states, could fall into the category of unsure, at best.

Alternatives to be considered in the future for waste processing and disposal include expansion of existing incinerator capacity, shredding and baling, followed by either rail haul to landfill areas or transportation to designated ocean-disposal locations [31], and development of new landfill locations in the area surrounding Boston.

SUMMARY AND CONCLUSIONS

The preceding information was presented to illustrate the magnitude of the expenditures required to operate the solid-waste-management system in Boston.

[19] Food-waste grinders generally have a negligible effect on sewer flow and blockage. However, there is a potential contribution to water pollution, especially in combined storm and sanitary sewer systems such as that which serves much of Boston.

Total expenditures for refuse collection, street sweeping, and lot-and-alley cleaning average $12.80 per person, per year. Refuse collection and street sweeping account for $7.13 and $5.33 respectively, or 97 percent of this cost. Costs associated with the operation of disposal facilities account for an additional $2.00 to $3.00 per person annually. These expenditures are comparable to those listed in the National Solid Waste Survey: $5.60 and $2.17 per person annually for collection and disposal.

The current level of street-sweeping allocations appears to be rather high when evaluated as a percentage of the total waste-management budget. This operation is highly labor-intensive (80 percent of budget) and a reallocation of current resources toward a more mechanized operation would probably reduce the cost, while equalling or improving the performance.

The lot-and-alley cleaning programs appear to be incapable (at current levels of financing) of providing the frequency of service and quality of maintenance required to maintain areas under their jurisdiction free of accumulated solid waste. The current allocations for these operations are small, at best, and expenditures should be increased to provide for better servicing of open areas within the city.

There are many areas in the waste-management system where new inputs could improve environmental quality in densely populated locations. Included among these would be modification of the current role played by each participant in the refuse-collection system; increased enforcement of the sanitary code relating to proper refuse preparation and collection; possible utilization of some on-site processing devices; and attempts at better integration of refuse collection and street sweeping such that any material escaping collection can be retrieved before becoming dispersed within the immediate neighborhood.

The existing system might be interrupted or modified at several points in order to achieve overall improvements. The inconvenience imposed on any of the participants (original producer, collection agency, administrator, and so on) through modification of existing conditions is difficult to assess. Modification of the action of the original producer and others involved in the management—or mismanagement—of wastes would improve the system considerably. The present operational sequence for removing refuse from curb or alley collection points appears to be adequate, though perhaps somewhat less than optimal. Material that is reasonably contained, though perhaps not in strict accord with the code, is transferred into the collection vehicle, with only little spillage. It is not clear from this study what the optimal allocation of effort is among education, inspection, enforcement, etc. in order to effect desirable modifications in the behavior of individuals.

The current allocation of inspectors for supervising collection personnel and soliciting compliance by residents with the Boston Sanitary Code appears insufficient. Of the two, the supervision of contract collectors (27 inspectors) appears to be more successful. The contractors and inspectors cooperate with each other to some extent. For example, the inspector will ensure that cars blocking alleys are removed by police tow trucks to speed collection; and in turn the contractor must return to a portion of the route to collet material that was not picked up.[20]

It is not clear how well the uniformed inspectors (12) operate in maintaining compliance with the Boston Sanitary Code. This group appears to be somewhat understaffed for policing the entire city adequately.[21] The predominance of mismanaged wastes in some areas indicates that modified enforcement policies are desirable.

---

[20] Occasionally such materials were placed out after the trucks had passed by.

[21] It should be noted that this organization has had a poor record for obtaining court convictions regarding violations of the Sanitary Code. A possible explanation for this is that until recently these offenses have been tried in District Court as

The potential for utilization of on-site processing equipment has been explored by the authors elsewhere [32]. The utilization of on-site processing equipment may necessitate development of programs for consumer education, extensive rehabilitation of structures, and large financial input.

Currently there is an attempt at coordinating refuse collection and street sweeping. However, examination of some areas within Boston discloses that the goals of this operation are not being reached.

Often much of the residue that escapes collection becomes widely dispersed before the opportunity for it to be collected by sweepers arises. An alternative to the present system would be to have street sweeping on the same day as refuse collection.

It is not unreasonable to request that portions of street-sweeping allocations be expended to maintain alley locations within the city. Mechanical sweepers cannot maneuver within the close confines of alleys, and additional expenditures for mechanical equipment to be employed in the streets would release some sweeping crews for deployment to alley locations, where they might be better utilized.

Some considerations should be given to a proposal which would enable a single contractor to provide complete sanitary service to a district. This situation currently exists in Charlestown, where the William Banfield Corporation provides both refuse collection and street sweeping at a cost of $7.15 per person per year compared with an average cost for the other districts of $12.80 per person annually. The motive behind this transition is obvious: such service would probably be more economical and efficient because the contractor would be more familiar with his service area; could better coordinate the total service system; and be more responsive to the requests of the residents within his service area. Also, there could be no instances where blame for poor waste management could be shifted from one service group to another. One contractor would provide all services, and alone retain responsibility for their results.

FUTURE INVESTIGATION

This initial study has established the present operating procedures employed by the City of Boston to control solid-waste material.

No rigorous method has been developed to establish the baseline conditions which define a tolerable environmental condition. The data required to develop these baselines are now available and the next phase of the study involves design of these parameters.

Deviations from this baseline will be evaluated in terms of the cost required to return the respective area to its baseline condition. The technological apparatus available for utilization in this maintenance program can be evaluated in terms of cost and production functions. Several alternative technologies must be evaluated to obtain the mix which yields the best return as measured in terms of system improvement, while imposing certain economic constraints.

Field monitoring has helped to explain the movement of uncollected solid waste within the environment. This information, coupled with the performance functions describing the productivity of the various municipal service groups, may reduce the maintenance allocation to a queueing-theory problem. The solution to this problem will result in a set of operating rules or guidelines which then can be selectively applied to certain areas, based upon the present state of the system with respect to its baseline condition. No such operating rules exist at present,

---

felonies, and judges have seemed hesitant to convict offenders. This poor record may be improved with the recent establishment of a city housing court designed to hear both housing and sanitary code violations.

and their value should be immeasurable in terms of improving the allocation of expenditures within a given maintenance system.

No statement can be made regarding the cost of a modified maintenance system, but in all probability it will not be cheaper than the present ineffective opertion. Returns from instituting such an operation will accrue in the form of more efficient maintenance systems, improved environmental conditions within the city, an increase in technically oriented employment opportunities (currently most agencies are faced with a shortage of unskilled labor), and the return of the inner city to a condition which is acceptable or even desirable.

REFERENCES

1. W. Z. Hirsch, "Cost Functions of an Urban Government service: refuse collection" *Review of Economics and Statistics, 47* (1), 87-92, 1965.
2. J. E. Quon, A. Charnes, and S. J. Wersan, "Simulation and Analyses of a Refuse-Collection System" *Journal of the Sanitary Engineering Division,* Proc. ASCE, 91 (SA5); 17-36, 1965.
3. H. B. Wolfe, and R. E. Zinn, "Systems Analysis of Solid-Waste Disposal Problem" *Public Works* 98 (9), p. 99-102, 1967.
4. M. M. Truitt, J. C. Liebman, and C. W. Kruse, "Simulation Model of Urban Refuse Collection" *Journal of the Sanitary Engineering Division* Proc. ASCE, 95(SA2): 289-298, 1969.
5. A systems study of solid-waste management in the Fresno area; final report on a solid-waste-management demonstration. Aerojet-General Corporation. Public Health Service Publication No. 1959. Washington, U.S. Government Printing Office, 1969, p. 411.
6. Annual report of Boston Public Works Department. Sanitary Division, 1968.
7. Personal communication. Leon Vignaux, Sanitary Division of Public Works Department, Boston, 1970.
8. Personal communication. Model Cities Office, Roxbury, Massachusetts, 1970.
9. *Manual of street-sweeping practice.* American Public Works Association, Chicago, Illinois, 1966.
10. Refuse-collection practice. American Public Works Association, Chicago, Illinois, 1966.
11. A study of solid-waste collection systems comparing one-man with multi-man crews, Report SW-9c, Contract PH 86-67-248, Ralph Stone and Company, Inc., Los Angeles, California, 1969.
12. Committee on Solid Wastes, American Public Works Association, *Refuse-collection practice,* 3rd Edition, Public Administration Service, Chicago, Illinois, 1966, "Analysis and Planning of Refuse Collection Systems," Appendix C, p. 473-503.
13. R. J. Black, A. J. Muhich, A. J. Klee, H. L. Hickman, Jr., and R. D. Vaughan, *The national solid-wastes survey, an interim report,* presented at the 1968 Annual Meeting of the Institute for Solid Waste of the American Public Works Association, Miami Beach, Florida, October 24, 1968.
14. Personal communication. Robert Bowes, Highway Department, City of Boston, September, 1970.
15. Public Works Report, 1967. Sanitary Division, City of Boston, 1967.
16. J. Flaherty, Boston's incinerator is a steam producer, *The American City, 29,* 104-106, June 1960.
17. J. R. Anderson and J. H. Dornbush, "Influence of Sanitary Landfill on Ground-Water quality" J. AWWA 59, p. 457-470, 1967.
18. "Law Concerning the Assignment of Places for Dumps and Incinerators" General Law Section 150 A, Chapter 11, Effective July 16, 1969.
19. M. W. First, "Waste Incineration at Sea and Ocean Disposal of Non-Floatable residues" Presented at the 62nd. Annual Meeting of the Air-Pollution-Control Association, N.Y.C.
20. Pollution control and management. Report prepared for New England Regional Commission by Charles River Associates, 1969.
21. Solid-waste disposal in Metropolitan Boston. Report prepared by Metropolitan Area Planning Council, 1967.
22. Automobile disposal, a national problem. Report prepared by U.S. Department of Interior and Bureau of Mines, 1967.
23. Personal communication. Leon Vignaux of Sanitary Division, Boston Public Works Department, 1970.

24. Personal communication. David Standley of Boston Air Pollution Control Commission, July, 1969.
25. Personal communication. Walter Smith of State Department of Public Health, September, 1970.
26. Personal communication. Boston Air Pollution Control Commission, September, 1970.
27. Economics of upgrading or discontinuing incineration, City of New York, Department of Air Pollution Control, New York, 1968.
28. Regulations for the control of air pollution in the Metropolitan Boston air-pollution-control district, General Law Section 142 D, Chapter III, Effective July, 1970.
29. Personal communication. Water Division of Metropolitan District Commission, September, 1970.
30. Personal communication. Walter Smith, Massachusetts Department of Public Health, Air-Pollution-Control Division, September, 1970.
31. Rail transport of solid waste – a feasibility study, Interim Report, U.S. Department of Health, Education and Welfare, Public Health Service, Bureau of Solid Waste Management, Cincinnati, Ohio, 1969.
32. L. J. Partridge, and J. J. Harrington, "Survey of Existing Apparatus for Handling and Processing Solid Waste at its Source of Generation." Unpublished Report, Harvard University, Cambridge 1970.

CHAPTER 5

OPERATIONS RESEARCH AND SOLID-WASTE COLLECTION

SUMMARY

This chapter deals with the application of a group of mathematical tools to the systematic analysis of the solid-waste-collection problem. The purpose of such work is twofold: to model the system in order to show its sensitivity to various parameters, and to predict the outcome of possible policy changes so that alternative management schemes may be evaluated. Ideally the analyst could build a simple model of the solid-waste-collection system and learn quickly all there is to know about its properties and optimal control. Such is not the case here. A combination of a very complex, interactive, technical, social, political and economic system to be approached, and the relative newness and crudity of the techniques available to the analyst, tend to make the approaches discussed here more first-cut screenings than truly powerful tools. This is not to imply that the use of such analytical techniques is a hopeless exercise. Presently, decision makers must manage a three-billion-dollar, 125-million-ton-a-year collection system essentially in the dark with little useful data or analysis input. Any additional information can only help to make that task easier and more efficiently executed.

GOALS

How then would the analyst approach the seemingly insurmountable problem of analysis for the decision-making process to be reasonably tractable? First the goals and objectives of the system must be clearly understood. As would be expected, this is a large-scale public-sector problem, and a general statement of objectives, such as "to move wastes from source to disposal with the greatest efficiency and satisfaction" is difficult to narrow down to definite guidelines. The benefits to be derived from a particular management alternative are largely of an aesthetic and public-health nature which are difficult to quantify and formalize. The costs of the system are not always in dollar terms but in social costs, such as changing land values, as well. This makes the process of evaluating the efficiency of alternatives quite difficult. Further, means for describing what level of satisfaction is to be gained by a particular set of alternatives is also vague. One might choose as measures of effectiveness such indicators as the degree of service offered in terms of frequency of collection, type of wastes removed, or location from which the waste is collected, or more amorphous measures such as street cleanliness or expressed customer satisfaction.

The usual approach to the problem, thus, cannot be direct in terms of balancing the benefits, costs and effectiveness but must be made by looking at some less-revealing but easier-to-solve subproblem. This will take the form of choosing control alternatives that will achieve some desired level of service and satisfy other constraints in the system at least cost. Recall that such an exercise must be carried out in the context of a large on-going system where attempts to experiment directly will not only be expensive but could well be catastrophic. Further, the available useful data base is virtually non-existent. The political implications of who controls the system or what group will gain or lose power by a change or who pays for which service may be the most persuasive argument for throwing out a scheme that might otherwise seem efficient. Since large investment has already been made in attempting to manage the system, the designer is denied the luxury of starting from the beginning and is often saddled with working around the blunders of the past. To this morass the analyst hopes to bring some order.

He basically is concerned with answering a set of questions about the system so that he can better understand it. The issues he might want to address are the following:

1. What level and type of service should be offered by the system? How will changing these services affect cost?
2. What type of vehicle should be used, and how many?
3. How many personnel are needed and what should their duties and work rules be?
4. What route should be assigned to each vehicle? How should the city be divided into sub-groups?
5. Are there parameters of the system to which system costs and variables are particularly sensitive?
6. Should there be intermediate transfer stations where collection vehicles could discharge their loads to vehicles better suited to long-distance transportation? Where should they be located and what type of equipment should they contain? What type of vehicle and how many should be used for transport to final disposal?
7. What facilities should be chosen as the best disposal alternatives and where should they be located?
8. What would be the effect on the system of new technology in waste reduction at the source, collection and/or disposal?
9. How will the stochastic nature of waste generation affect the analysis? How will the solution change as the area to be served continues to grow and expand?
10. What are the effects of political, social, and economic constraints? How much should be spent on aesthetic factors? Is regional grouping a feasible alternative?

COMPROMISE SOLUTIONS

Ideally, to answer all these questions, the analyst would like to build a model that would be able to follow the collection process in its minutest detail and be able to manipulate all of the myriad parameters that could possibly affect the solution. However, its magnitude and complexity are such that to consider the system in its entirety and to encompass every possible detail is quite difficult, if not impossible. For this reason, simplifying assumptions must be made in model development which will allow some approximation but not the fine details, of the problem. Basically, two different approaches are taken. If one is willing to neglect the problems of routing of the individual vehicles through their collection tasks, the large-scale problem of how material should be moved through the system and how transfer-facility location may be chosen can be approached optimally. This will be called the "micro-scale" or flow-of-materials subproblem. If the location of facilities and collection areas assigned are known for a set of collection vehicles, the small-scale problems of routing vehicles among individual tasks may be approached. This second problem will be referred to as the micro-scale problem. Developmental work and application to actual solid-waste-collection systems has been done and is reported in the next two sections.

THE MICRO-SCALE SUBPROBLEM

In this section, several different approaches to the problem of modeling the collection process in great detail in terms of its basic parameters are discussed. The most detailed are the simulation models which appeared first on the scene and have had the most extensive use. These models are characterized, however, by specific general collection areas rather than actual route assignments. Application of such models to the Baltimore, Maryland; Winnetka, Illinois; Chicago, Illinois and Los Angeles, California systems are reported. Concurrently, work in the general area of routing and scheduling vehicles among tasks has developed and this work, along with its limited application in solid wastes in Potsdam, New York and Cambridge, Massachusetts, is discussed. Finally a summary of the basic data and of the variable-specification problems for such studies is presented.

Simulation Modeling

The first detailed attempt at simulation modeling of the interdependence of the significant parameters in the waste-collection system was reported by Quon, Charnes and Werson [1]. Based on data from the Winnetka, Illinois system, a model was built to simulate the operation of collection vehicles with collection speeds and waste loads generated randomly from distributions observed in the Winnetka system. Two basic work rules exist for job assignment in solid-waste collection. One is the daily-route method in which a truck is given a specific assignment to complete. When this is completed, the crew may go home with pay for an eight-hour day, even if they finish sooner. Routes are usually assigned so that no overtime is needed and must be fairly equal in work content so that employees will feel they are treated equally. The other work rule is the constant-length-day rule which has the crew attempt to collect as much as possible in eight hours. Such a scheme requires more supervision and makes work fairly difficult to schedule. Most cities use the daily-route scheme (often referred to as the incentive method) for this reason. However, considerable care must be taken in establishing routes. Winnetka uses the daily-route method so the simulation included this policy as well as the policy of using a driver and two crew members and twice-a-week collection frequency.

Input parameters for the simulation were as follows.

1. Average refuse produced in pounds per service per day where a service is defined as one stop of the collection vehicle.
2. Standard deviation of refuse production.
3. Number of services in each area.
4. Days of the week for collection from the area.
5. Internal distances in the collection area.
6. Distance from the collection area to the disposal site and to the garage.
7. Truck capacity.
8. Number of collection services assigned per truck.
9. Equation form for computing pickup time.

The methods for computing pickup time (a regression equation based on characteristics of the system) and waste loads (generated randomly from distributions) are reported in the next section. Truck capacities were three and five tons and the simulator could handle up to

1,000 collection services and 64 trucks. As noted, the assignment of trucks to collection areas is given as input data and no routing of the vehicles in the collection area is attempted.

Results from this study of Winnetka, Illinois indicated:

1. A relative insensitivity in collection to the stochastic variation of wastes generated in lower ranges of variation. As variation becomes higher, the daily-route method loses its efficiency.
2. Partial loads (trips to disposal when significantly less than full because of time or collection-area constraints) are difficult to avoid by changing assignments.
3. Truck capacity of from 1/2 to 1/3 of average daily collection assignment gives the best efficiency.
4. From queueing studies at the disposal site, an unloading dock for every three to four vehicles is needed.

In a later extension of their model to Chicago, Illinois, Quon, Tanaka and Wersan [2] made significant changes to deal with a system with constant-length-day work rule. They added the following policy alternatives.

1. Allowance for overtime pay.
2. The possibility of using a relay driver to take the collection vehicle to disposal after working hours which would allow the actual crew more time in the collection areas.
3. Different assignments in terms of the refuse quantity expected in number of truck loads per day and expected length of work day for each collection crew.
4. Varied frequency of service to be provided.

Chicago work rules – a constant-length work day and once-a-week collection frequency – were used as invariants as well as input data of the type described for the earlier study.

Investigations of the effect of using a policy of relay drivers without overtime showed significant increase in service in terms of number of services collected on time with only slight additional cost. Use of overtime to collect only loads scheduled for collection that day which might otherwise be delayed showed the same result. However, neither policy change could compensate for too large an assignment.

Truitt, Liebman, and Kruse [3] built a simulator to investigate system response to changes in operating policy measured in cost per ton of collected refuse. They looked at two policy areas: the use of transfer stations and increases in collection frequency. They simulated trucks moving from a disposal site to a collection area, through that area and back to disposal. No attempt was made at routing the vehicle through the area, and collection times were based on average collection speeds and times for a given waste load to be collected, all of which were generated randomly from observed system histograms. Since they planned to look at the Baltimore system, the basic work-rule policy was to assign given routes rather than a constant-length work day. Provision in the model for the location of one given transfer site was also included. The basic Baltimore work rules of a driver and a crew of three for collection in areas which had not been serviced in four days and a driver and a crew of two in areas which had not been serviced in three days were used as well as equipment with five-tons (20-cubic-yard) capacity.

The data needed for the simulation were these.

1. Boundaries of the area, and of the subareas with their population-density classifications.
2. Frequency of collection.
3. Collection-truck capacity.

4. Transfer-station location (if any).
5. Size and type of transfer trailers (if any).
6. Location of final disposal site.
7. Labor pay scales, including overtime rate.
8. Probability distribution functions for collection rates, unloading times, truck speeds and amount of waste generated per capita per day.
9. Average number of persons per housing unit for each neighborhood density classification.
10. Amortization interest rate and time period for transfer-station structures and equipment.
11. Capital investment in transfer-station land.
12. Hourly operating cost for collection trucks including amoritization; maintenance and repair; and operation (but excluding driver and crew wages).
13. Hourly operating cost for tractor-trailer rigs as in item 12.
14. Monthly cost of utilities for the transfer station.

Each simulation run was for one week and the program calculated the number of trucks needed and assigned areas to each of these trucks for each day.

The northwest quadrant of the City of Baltimore, Maryland, with 75,000 housing units and 225,000 persons, was chosen for study with the model. After verification runs, a set of experiments were run to show the cost of increasing collection frequency based on the assumption that collection time per ton would increase due to decreased loadings at each stop but that the total waste load collected would not change with frequency. As will be discussed in the next section, the work of Quon, et al. [4] on the relationship between waste generated and collection frequency may prove this to be an assumption that seriously biases the results.

The resulting increase in collection costs were about 9 percent with 60 percent of that amount due to equipment increases and 40 percent to labor. The investigators also noted that increasing collection frequency also increases the peak loads to the disposal facility.

Investigation of the possible use of a given transfer-station site showed that at the present haul distance (about eight miles average one way), the cost savings are marginal. A greater distance to the disposal point was necessary to justify building and operating a transfer facility. This simulator is now being used by the Baltimore City Department of Public Works for further investigation of its system.

Stone [5] in a study prepared for the United States Public Health Service, presents a detailed study of the mechanics of collection. Large-scale data collection from 234 cities throughout the United States and an intensive study of four municipality-run services and two private companies in California are reported. Some care should be taken in using these data, as the basic underlying assumptions for each service reported are not always defined. The intensive studies included time-and-motion studies and the use of video tape to record various activities. The whole study was geared to an investigation of crew size, in particular to the one-man crew (combination driver-collector). The main conclusions are reported in the next section. However, Stone did attempt some noncomputer mathematical-model studies where average figures were used throughout. In the report, distributions are generated for such measures of efficiency as collection time as a function of number of cans per stop; collection time as a function of the number of cans and crew size; and travel distance between stops and average collection times. Under some very specific conditions, the report found one-man collection crews to be most effective except for the cases where barrels had to be carried from backyards.

Route-generation Models

In this section, the basic question of how individual routes and assignments for crews may be specified is approached. The problem of vehicle routing is a fairly general problem in management and has received considerable attention in the past few years because of the obvious payoffs to be gained in increasing routing efficiency. However, the very large scale needed to deal with the problem in the context of an entire city involving thousands of streets immediately makes the task of generating good routes quite challenging. The problem is characterized by the fact that it is not too difficult to build a feasible route or set of routes which satisfy all necessary time, distance, and volume constraints. However, generating very good or even optimal routes is extremely difficult because the number of feasible routes increases exponentially with the number of streets. Attempts at generating both good feasible routes (heuristic methods)[1] and optimal routes will be presented with emphasis on those that have been applied to solid-waste problems.

One way of approaching the problem is to break areas into subareas involving very small collection tasks and then look at good ways of combining them into daily or weekly crew assignments. Coyle and Martin [6] presented a heuristic algorithm based on dynamic programming for this type of aggregation process. A sample problem based on a large English city indicated savings of almost a quarter-of-a-million dollars per year due to the efficiencies gained. It would seem that in view of the problems to be encountered in actual routing in large cities that approaches such as these (i.e., districting or aggregation techniques) might prove to be a fruitful research area.

Bodner, Cassell, and Andros [7] have built a simulation model which operates at the scale of the street network and generates feasible routes randomly. They consider such parameters as crew size, vehicle size and pickup time. Required for the simulation is a detailed representation of the street network which could severely limit a large problem. Waste loads and collection times are generated from distributions and the truck is routed by making a decision at the end of each street of whether to go to disposal because of a capacity restriction, to the garage because of a time constraint or randomly to choose another street in which to collect. With repeated application of the technique, a large number of feasible routes are generated from which the 'best,' based in their case on minimum distance traveled, is chosen. It should be noted that other criteria, such as minimum time, for good routes might also be applicable.

They chose as an example a small town (Potsdam, New York) which has a population of 11,000 and uses one truck for once-a-week collection. In the example, 100 runs were made with no stochastic variations of load and the minimum-distance route was chosen. There was a significant difference between the average route distance for the trials and the minimum-distance route selected. The investigators refer to this as the "optimum" route – which is not necessarily true since only a few alternatives have been searched. However, in a statistical sense, this size sample should produce a fairly good feasible route.

The chosen route was then subjected to a long-term simulation with stochastically varying waste loads. The authors considered the policy of using either a three-man or four-man crew (including driver) and 12- or 16-cubic-yard vehicles, and found a four-man crew with a 16-cubic-yard vehicle best for their case. However, since the size of the city under investigation was such that it could be entirely serviced by one 16-cubic-yard truck and a four-man crew in a week with no overtime, little of a general nature may be abstracted from their findings.

---

[1] "Heuristic" implies a good, feasible solution which is not necessarily optimal.

Looking at the general problem of routing vehicles, Marks and Liebman [8] presented a detailed description of the various techniques attempted in the literature. Basically vehicle-routing problems may be categorized by the responses to the following questions.

1. Are the services to be rendered by the vehicles located at a finite number of points which have given location coordinates (the discrete problem) or are they in essence continuous in that every street or set of streets must be covered (the continuous problem)?
2. Is vehicle capacity (defined both in terms of volume and weight) such that the vehicle can accomplish the entire servicing operation in one continuous route? If so, the problem is a single-route problem. If not, either more than one vehicle is needed or the single vehicle must return to its origin (such as the garage or disposal point) more than once. In either case, we are dealing with a multiroute problem.
3. Are there additional constraints on routing such as precedence, earliest and latest arrival time, and so forth?

Traveling-salesman Problem

The case of single-route discrete problems without additional constraints is known as the 'traveling-salesman' problem. Here the objective is to find the minimum-distance tour a salesman would follow to visit N cities (or points) and return to his origin so that all cities are visited exactly once and his tour is continuous. Distances between all city pairs are assumed known and may or may not be symmetric. For the asymmetric case, an N-city problem has (N-1)! feasible solutions, a number which very rapidly becomes large as N increases. Success in recent years has been limited to problems involving 100 cities or points, which are not large when applied to solid-waste collection. However, in this case we are dealing with multi-routes. This problem has been called the M-salesman traveling-salesman problem [8] and computational success with any fair-sized problem is extremely limited. Most work in this area deals with heuristic methods of establishing good feasible routes. The most easily available model for persons wishing to experiment is an IBM applications program available for the IBM 360 computer series. It is entitled the Vehicle-scheduling program [9] and is based on the heuristic methods developed by Clarke and Wright [10]. The program allows a fairly large and complicated street network, capacity constraints on vehicles, flow barriers throughout the network and other sophistications which make the package quite attractive. No attempt has been made to apply it to solid-waste problems so far as is known.

Chinese-postman Problem

One would certainly argue that viewing the solid-waste collection – routing problem as discrete rather than continuous may be greatly enlarging the problem. Until recently, the only answer to this has been that at least one could approach a solution of the discrete problem but had no way of dealing with the continuous problem. However, recent developments not only show us a hint of how to deal with continuous problems but indicate the ability to deal with much larger problems in a shorter time.

The single-route continuous problem is called the 'Chinese-postman' problem. It is defined as finding the minimum-distance continuous tour through a network which travels all arcs at least once. Thus the Chinese-postman problem is an arc-covering problem while the traveling-salesman problem is a node-covering problem. Recent developments in the Chinese-postman problem using graph theory point to quick and efficient methods of solution. Stricker [11], in a discussion of the application of these problems to solid-waste collection routing, describes the theoretical developments and some heuristic algorithms. He points out that in fact cities present multiroute problems which include the additional difficulty of one-way streets. His work contains a heuristic for finding routes by decomposing cities into more workable sub-areas. When applied to the Cambridge, Massachusetts system (population about 100,000 giving about 125 tons per day and using 12-15 collection vehicles) he was able to demonstrate considerable distance savings in routes over current practice.

From the discussion, one would suspect that the need for research in all these areas is necessary and would reap considerable benefits.

Data and Assumptions Used

In all analytical work, the data used and assumptions made are a basic part of the work and should be carefully noted. The following is a summary of the problems faced in specifying the various systems reported here and the way that investigators dealt with them in addition to some comments on prevalent practice.

Amount and Variability of Waste Generated

A key factor in the analysis of solid-waste collection is that the waste generated per person per unit time is a stochastic variable and all of the simulation models reflect this. Three general problems are apparent.

1. What is the mean, μ, and standard deviation, σ, of waste generation in the collection system?
2. Is there a significant trend in the data for waste production over time?
3. How do waste loads change as system policy changes?

Bodner, et al. [7] in studying once-a-week collection in a small town reported an average of 56 pounds per week per service and uses values of from 0.1 to 0.6 for the coefficient of variability, (σ/μ). Truitt, et. al. [12] reported a mean of 1.95 pounds per day per person and $\sigma = .046$ for purely residential, twice-a-week collection in Baltimore. Quon et. al. [4] in his work in Winnetka, Illinois where twice-a-week collection is used reported $\mu = 79$ pounds per service per week with (σ/μ) varying from 0.07 to 0.26.

The wide variation here is typical of almost all reports of field data. Part of the differences is due to the "apples and oranges" effect caused by attempting to compare conditions measured under different assumptions and conditions. Each city has evolved an individual system in terms of the types of waste collected, the frequency of collection, where it is collected from, the methods used for financing and charging, and even the types of data reported. Thus, care should be taken in trying to compare services on the basis of reported data until these differences are accounted for. If possible, in the analysis of a given system, actual field data should be collected.

The most important work concerning waste generation and collection policy is Quon, et. al. [4] in which a detailed statistical analysis on two wards in the Chicago, Illinois system was presented. Here once-a-week collection existed and experiments were made for a change to twice-a-week collection. The authors noted a seasonal variation in waste generation with the lowest amount in winter, greater-than-average in spring and above average in the fall. Trends from 1950 to 1960 showed an almost 30-percent increase in waste generated for that time period but in the years 1961 to 1966 the waste is saved for collection under twice-a-week frequency. Solid-waste collection is part of a larger-scale waste-accumulation-

and-transportation system which includes street cleaning, bulky-item collection, vacant-lot cleaning and the maintenance of sewerage and drainage catchbasins. The additional waste loads may well represent a decrease in the cost of cleaning streets, neighborhoods, and sewers as well as a reduction in air pollution due to burning.

In any event, all attempts to change policy in the system must be guided by this important example. There is no doubt that certain collection policies do change the amount of waste to be collected by the service.

Pick-up Time

The question of how to deal with the estimation of pick-up time in the analysis has been dealt with quite differently by each of the investigators. It is clear that such factors as crew size, days since last collection, location of refuse to be collected, type of container, equipment size, weather conditions and other work rules all may affect performance. Quon et. al. [13, 14] used a regression equation based on fitting of data in the Winnetka and Chicago, Illinois systems. They hypothesised a relationship of the following form which gives the time spent in collection.

$$\bar{\theta} = \alpha S + \gamma m_i$$

where $\bar{\theta}$ ≡ the average pick-up time for S services in minutes

S ≡ the number of services

α ≡ constant, characteristic of the time needed to collect refuse from a service, in minutes per service (obtained from system data)

γ ≡ constant, characteristic of the time spent in travel within the collection unit

m_i ≡ internal mileage traveled per trip in miles

$= S/\rho_s$

ρ_s ≡ service density – number of services per mile

From a correlation analysis in Winnetka, with a crew size of a driver and two laborers and twice-a-week collection, Quon found:

$$\alpha = 0.717$$
$$\gamma = 33.1$$

Truitt et al. [12] chose to deal with average collection rates in pounds per hour as a function of neighborhood density and days since last collection. Through statistical data, they showed that only two density types need be considered with ten housing units per acre being the dividing line. Crew sizes of four men early in the week and three men later in the week was fixed to fit Baltimore policy. The investigators presented observed histograms of pounds per hour collected versus frequency for Baltimore in the following cases.

1. A density of less than ten houses per acre and three days since last collection.
2. A density of less than ten houses per acre and four days since last collection.
3. A density of greater than ten houses per acre and three days since last collection.
4. A density of greater than ten houses per acre and four days since last collection.

Assumed histograms for both density types under an assumption of two days since last collection are also presented. The simulator then samples from these distributions randomly and thus incorporates collection and between- collection time.

Stone [5] presented histograms for collection time only, with different crew sizes and different number of cans per service, in California. Differences due to using disposable bags rather than cans were documented. Average collection times based on observation were also reported. The Stone report should prove to be a valuable data source for all those interested in analysis in this area.

Crew Size, Type of Vehicle and Work Rules

Reported crew sizes in different systems vary from one to nine men with very different types of equipment and work rules. The first important work rule concerns whether or not the driver is involved in the collection practice. In most east-coast cities, for example Baltimore, Maryland; New York, New York; and Cambridge, Massachusetts, crews range from two to four men in addition to the driver, who never leaves the cab. Stone [5], in work on the west coast, counted the driver as an active part of the work crew as is customary in practice there. He made a detailed time-and-motion study to find that a one-man driver-collector is more efficient than a two-man or three-man crew. The practices he outlined would be strongly opposed by the various public-employees units if an attempt were made to employ them in systems where they do not now exist.

Another difference is the collection location. A majority of collections are at curbside or from alleys where the customer has placed his load. In Cambridge, Massachusetts, however, 'barrel rollers' precede the trucks in some areas to bring the waste from the backyard. The type of container and its maximum loaded weight also varies. Experiments with disposable plastic bags instead of cans indicate considerable collection efficiency if the bags are strong enough to keep from tearing during the collection process and if animals cannot get at the bags waiting for collection.

The specific types of work assignment such as constant routes or constant-length work days have been discussed earlier. Most major cities use an incentive or constant-route plan which allows crews to leave work with a full day's pay whenever the work assigned is finished. This results in better work efficiency, customer service, and requires less supervision if the routes are properly designed to ensure equal work loads.

The type of vehicle used differs in reported size for the main collection vehicle from 12 to 35 cubic yards. The larger vehicles are more predominant in the west where streets are wider and straighter. The types of compaction equipment available are discussed in Chapter 3.

Travel and Waiting Times

Several types of travel time are of interest: time between collection; time to move from the collection area to disposal for collection vehicles; waiting times at the disposal point; time to return to the garage; as well as time for compaction system on the vehicle to operate and for large-scale transport vehicles to operate between intermediate points and final disposal. As noted, Truitt and Quon avoided defining a "between-collection" time by lumping it into an overall collection rate. They did, however, present estimates for truck travel speed outside the collection area and the waiting time at disposal. Stone [5] presented a frequency distribution for travel time between collection stops as well as other speed estimates for the system.

MACRO-SCALE MODELS

Macro-scale models are distinguished by the fact that they ignore individual routing problems in collection areas but instead concentrate on how the waste material flows from its source to disposal points. In this context, specific additions to the network of alternatives such as transfer stations, new types of transport, and new disposal sites and methods may be evaluated. The most general and far reaching of the systems studies carried out in this context is one still under way (in 1970) at the University of California at Berkeley entitled "Comprehensive Studies of Solid-waste Management." In their annual reports [Golueke and McGauhey, 15, 16] one of the stated objectives of the study is "to explore the potential of operations research to help in the definition and solution of the solid-waste-disposal problem.

Basically, two problems were investigated. The first was to find a means for evaluating solutions for the total refuse-disposal problem that would not only be applicable to a wide spectrum of alternatives but would be sensitive to environmental and governmental constraints as well. This involved the development of conceptual models which gave insight into the operation of the system. As a second problem, the study undertook to build mathematical models for the conceptual models. These included a waste-generation model, a waste-collection, treatment and disposal model, a regional economic model and models for population, public-health aspects, land use and process technology. The most interesting of these models from the viewpoint of operations-research techniques was the waste-collection, treatment and disposal model. The subgroup working with this problem looked at the means of determining how solid wastes should move from sources through treatment and processing to disposal sinks. They identified the problem as one of network analysis and suggested building mathematical models to solve the process. The outcome of the development of specific techniques for solving the model as formulated was reported by Anderson [17], who was one of the study group. Anderson was interested in modeling how the flow would be optimized through a given system which contained existing facilities and potential new facilities. He identified the problem as a study of a trans-shipment network with some peculiarities due to the nature of the processes involved. In considering a total waste-management model, including intermediate treatment facilities, it is necessary to consider the changes in volumes and in the state of the wastes; these changes cannot be treated in ordinary network algorithms. In his models all costs were linear, and the system under investigation considered the gross shipment between points and not the routing of individual vehicles among small collection tasks. Also, Anderson dealt with the problem of finding flows through a known or given configuration. To consider various combinations of new facilities, each scheme must be evaluated separately. No methodology was presented for a systematic search over facility-location possibilities. The assumption of linear costs omitted some of the important aspects of the problem. Although nonlinear costs may be given linear approximations, this procedure is valid only if the nonlinear costs show increasing marginal cost. Facilities costs are commonly concave or quasi-concave – that is they show decreasing marginal costs or "economies of scale." Thus linear approximations either will not be successful or must be done in such a manner that one loses confidence in the meaning of the results of the solution.

Other studies defining the solid-waste system have been carried out by consulting organizations for governmental units, but few if any have gone beyond the characterization of the system and attempts at simulation of the process. Such studies include reports by Management Technology, Inc. [18] for the State of Maryland, and by Aerojet General [19] for the State of California.

Others who have worked in the macro-scale models have approached the problem as the optimal location of intermediate facilities. This problem is viewed as a tradeoff between the savings in transportation cost to be gained by additional facility costs. Skelly [23] presented such a model in his Ph.D. thesis which is based on a heuristic solution technique. Since the value of such models is not in running them once but in using them under a variety of different conditions to show the sensitivity of the parameter changes, non-optimal approaches may be deceiving when comparisons are made between runs. Apparent sensitivity, or lack of it, to a particular parameter may be due to the solution technique rather than the parameter change. Marks and Liebman [8, 20] and Marks, Liebman and Bellmore [21] reported on an optimal technique and its actual application to the problem of locating transfer stations in Baltimore. The same northwest quadrant of the city as reported on by Truitt et. al. [3] was chosen for investigation. Six possible sites for transfer to road vehicles and one rail-haul site were given as possible alternatives. The road vehicles were considered to be large tractor trailers capable of handling 30 to 35 tons per trip. Rail haul to Western Pennsylvania from the central rail yards was also estimated. Investigations of whether or not the decision to build transfer stations was feasible were carried out under a variety of different assumed conditions. The immediate results of this study using readily available data for screening purposes were as follows.

1. At the present time, because of the configuration of the system with disposal located fairly close to the collection areas (eight miles average), transfer was feasible but would represent only a five to eight percent saving. Further, under almost all conditions, only one transfer station (transfer to road vehicles) would be built. It would be in the capacity range of 900 to 1,000 tons per week. See Figure 5.1.
2. Investigations of how the decision would change as loadings and haul distance increase showed the transfer alternative to become more favorable over time as load locations and disposal sites become more dispersed due to population growth and spread. See Figure 5.2.
3. The raul-haul alternative would cost about 12 percent more to operate than the present

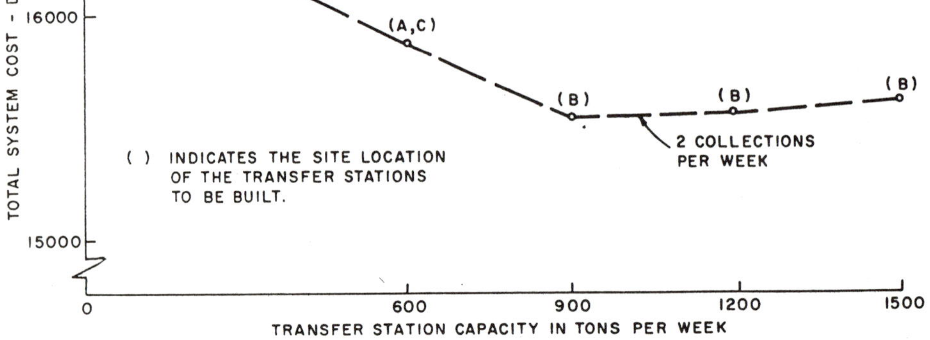

Figure 5.1 Plot of total system cost versus size of transfer station.

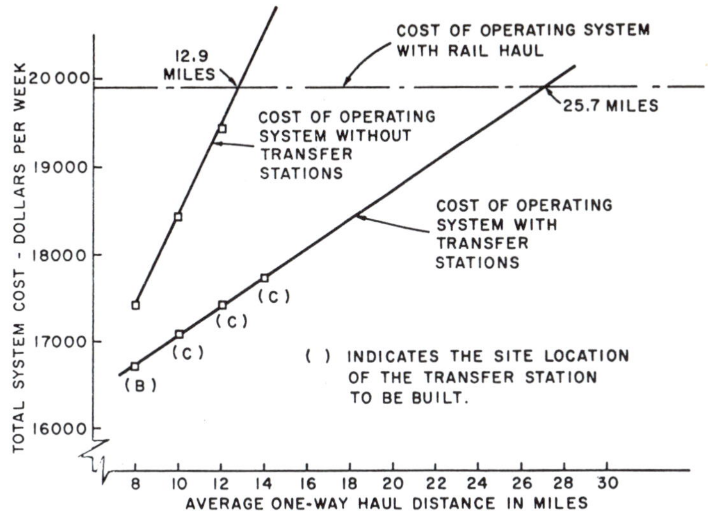

Figure 5.2 Plot of average haul distance versus cost of total system for three-times-per-week collection frequency.

system and about 18 percent more than the system with a vehicle-transfer facility. Haul distance to disposal by collection vehicles would have to increase an average of from 3.5 to 5 miles for rail haul to be favorable without transfer and from 14 to 18 miles to be favorable with transfer.

4. The cost of changing the frequency to two times a week from three times a week collection is difficult to estimate because of the sensitivity of this calculation to estimates of how much the waste load would be with three-times-a-week collection. Estimates of the added cost range from 4.5 percent with no increase in total waste load to 21.5 percent with a 15 percent increase in waste load. All of these figures are based on the assumption that routes for vehicles collecting three times a week be redesigned for efficiency purposes. As noted earlier, the estimate of waste loadings under different conditions is the weakest link in the analysis process.
5. While differences in assumptions make estimating the cost of some programs difficult, the site selected for a transfer facility is remarkably insensitive to parameter change. Regardless of the collection frequency or collection routes the same site is always chosen for the transfer facilities (Figure 5.3). If for some reason, the best site could not be used, two other sites appeared as good alternatives and it would cost little to shift to them. This gives the decision maker some latitude if one site is more difficult to obtain because of political pressures or other social factors. Similarily the efficiency to be gained by regional cooperation may be shown and evaluated.

Another investigator approached the problem of optimal operation of solid-waste-treatment facilities. Stern [22] looked at the problem as a trade off between holding or storage costs and processing costs. He presented a queueing model with stochastic arrival stream of raw waste, a processing operation with variable rates and a storage rate. The objective was to define the service-rate rule that minimizes the long-term expected value of total discounted cost. Cost was the sum of processing and holding costs. Proof of optimal solution for the models presented were shown but no applied example in solid waste was carried out.

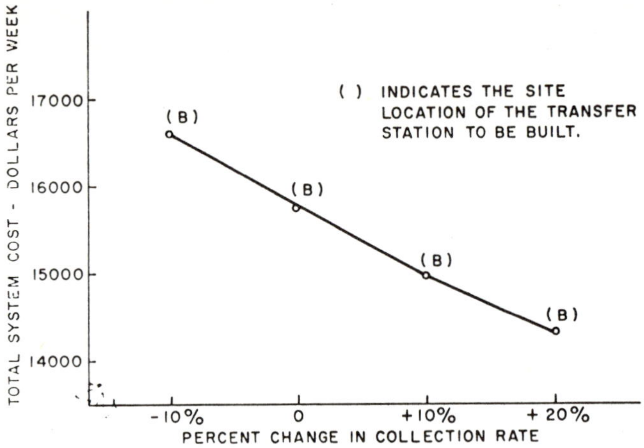

Figure 5.3 Plot of varying collection rate versus cost of solution.

CONCLUSION

Operations-research analysis in routing collection vehicles and in siting solid-waste-treatment facilities has been shown to yield substantial gains in efficiency in many situations. As better data become available and as improved techniques are developed, the potential gains may be expected to be increased.

REFERENCES

1. J. E. Quon, A. Charnes, and S. Wersan, "Simulation Model of a Refuse-collection System" *Journal of the Sanitary Engineering Division* of the American Society of Civil Engineers, Volume 9, SA 5, October, 1965.
2. J. E. Quon, M. Tanaka, and S. Wersan, "Simulation Model of Refuse-collection Policies" *Journal of the Sanitary Engineering Division,* American Society of Civil Engineers, Volume 95, SA 3, June, 1969.
3. M. M. Truitt, J. C. Liebman, and C. W. Kruse, "Simulation Model of Urban Refuse Collection" *Journal of the Sanitary Engineering Division,* American Society of Civil Engineers, Volume 95, SA 2, April, 1961.
4. J. E. Quon, M. Tanaka, and A. Charnes, "Refuse Quantities and Frequency of Service" *Journal of the Sanitary Engineering Division,* American Society of Civil Engineers, Volume 94, SA 2, April, 1968.
5. R. Stone, "A Study of Solid-waste-collection Systems Comparing One-man with Multi-man Crews" final report by Ralph Stone & Company to the U.S. Department of Health, Education and Welfare, Public Health Service, Pub. 1892, 1969.
6. R. G. Coyle, and M. J. C. Martin, case study "The Cost Minimization of Refuse Collection Operation" Operations Research Society of America Meeting, May, 1968.
7. Richard M. Bodner, Alan E. Cassell, and P. J. Andros, "Optimal Routing of Refuse-collection Vehicles" *Journal of the Sanitary Engineering Division,* American Society of Civil Engineers, SA 4, August, 1970.
8. D. H. Marks and J. C. Liebman, "Mathematical Analysis of Solid-waste Collection" Department of Geography and Environmental Engineering, The Johns Hopkins University, March, 1970a.
9. Anonymous. System 360 vehicle scheduling program (360A-ST-06X) application description, International Business Machine Corporation, Technical Publications Department, White Plains, New York, 1968.
10. G. Clarke and J. W. Wright, "Scheduling of Vehicles from a Central Depot to a Number of Delivery Points" Operations Research, Volume 12, 1964.
11. R. Stricker, "Public-sector Routing Models: The Chinese-postman Problem" Masters Thesis, Operations Research Center, Massachusetts Institute of Technology, September, 1970.

12. M. M. Truitt, J. C. Liebman, and C. W. Kruse, "An Investigation of Solid-waste Collection Policies," Department of Environmental Health, The Johns Hopkins University, Baltimore, Maryland, 1968.
13. J. E. Quon, A. Charnes, and S. Wersan, "Simulation Model of a Refuse-collection System" *Journal of the Sanitary Engineering* Division of the American Society of Civil Engineers, Volume 9, SA 5, October, 1965.
14. J. E. Quon, M. Tanaka, and S. Wersan, "Simulation Model of Refuse-collection Policies" *Journal of the Sanitary Engineering Division,* American Society of Civil Engineers, Volume 95, SA 3, June, 1969.
15. C. G. Golueke, and P. H. McGauhey, "Comprehensive Studies of Solid-wastes Management," First annual report, Sanitary Engineering Research Laboratory, University of California at Berkeley, Report 67.7, May, 1967.
16. C. G. Golueke, and P. H. McGauhey, "Comprehensive Studies of Wastes Management," first annual report, Sanitary Engineering Research Laboratory, University of California, Report 69.1, January, 1969.
17. L. E. Anderson, "A Mathematical Model for the Optimization of a Wastes-management System" Sanitary Engineering Research Laboratory, University of California, Berkeley, Report 68-1, February, 1968.
18. Anonymous. "The applicability of a Systems Approach to Solid-waste-management Problems," report presented to the Maryland State Department of Health, Division of Solid Wastes by Management Technology, Inc., July, 1966.
19. Anonymous. "California Waste-management Study: report to the State of California, Department of Public Health" by Aerojet General Corporation, Azuza, California, 1965.
20. D. H. Marks, and J. C. Liebman, "Optimal Location of Solid-waste Transfer Stations," presented at the 37th National Meeting of the Operations Research Society of America, Washington, D.C., April, 1970.
21. D. H. Marks, J. C. Liebman, and M. Bellmore, "Optimal Location of Intermediate Points in a Transshipment Network," presented at the 37th National Meeting of the Operations Research Society of America, Washington, D. C., April, 1970.
22. H. I. Stern, "Optimum Service Policies for Solid-waste-treatment Facilities," Sanitary Engineering Research Laboratory, University of California, Berkeley, Report SERL 69-6, May, 1969.
23. M. Skelly, "Planning for Regional Refuse Systems," Ph.D. Thesis, Cornell University, September 1968.

CHAPTER 6

LANDFILL

The open, burning dump has become a symbol for all that is wrong with our treatment of solid wastes. Town dumps employ the nearest land of low economic value, and often this has meant filling a marsh or swamp. New interest in and information about our environment has shown that wetlands, both coastal and inland, play important roles in the life cycles of many organisms and animals which are vital links in the food chain of higher animals and of men. To fill wetlands with refuse may turn out to be an expensive error.

Accordingly, there is a public clamor for better methods of treatment of solid wastes, and part of that clamor has often been the almost automatic rejection of the use of any land for the disposition of our solid wastes. In many of our cities the existing landfills are full, and yet the selection of almost any new area for filling raises a public outcry. With a background of a frequently repeated statement in newspapers, magazines, and television that "we are burying ourselves in trash," the public wants engineers and scientists to produce something which causes trash simply to disappear.

At the present time, the nearest approaches to such a process are incineration and salvage. Neither of these processes violates the law of conservation of mass, which is so cheerfully broken by many commentators. Virtually all of our refuse originates in the ground, and the fact that it may accumulate in vast quantities in a central city is the result of a network of transportation and processing spreading out to remote regions of the country and of the world to places where materials are extracted in mines, quarries, farms, and forests. Since only a comparatively small fraction of the goods we produce are scrapped each year, the holes we are making in the ground are growing faster than are the piles of solid wastes in our cities. On a national and on a global scale, therefore, landfill in some form is very desirable. To fill an abandoned strip mine or sand pit can be beneficial from the environmental, ecologic and economic aspects.

Seen in this light, the so-called "solid-waste crisis" becomes merely one of transportation and handling. In this chapter we wish to review some methods which reduce the transportation and handling costs and produce landfills which are tolerable to the local communities during filling, and which are highly beneficial subsequently.

SANITARY LANDFILL METHODS

The term "sanitary landfill" has been given a fairly exact meaning in terms not only of the effects but also of the processes of landfilling. However, other methods of putting refuse in the land which do not result in any of the nuisances associated with open dumping have also been classified as "sanitary." We shall call the original system "basic sanitary landfill" and compare the apparent advantages and disadvantages of other methods with it.

Some of the methods described briefly below must be classified as being in a research or experimental stage at present. These methods deserve a place with the others because it seems likely that in certain circumstances their use will be profitable. For there is one conclusion which is easy to reach with regard to landfilling methods: no one method is best in all circumstances. The conditions where the use of each method will most likely be optimum are discussed.

Basic Sanitary Landfill

Refuse is dumped and compacted (by heavy steel-wheeled or tracked vehicles) in "cells" which may be from two to 15-feet deep (a maximum depth of ten feet after compaction has been recommended [1], but there is nothing absolute about this figure) and of a width and length determined by the quantity of refuse arriving in each day. At least at the end of each day, and sometimes more often than this, the cells are covered with a thin (six-inch to two-feet) layer of earth or other cover material (incinerator ash may be used if it is inert). The final cover for the landfill is two to six feet of inert fill or loam.

The principle of basic sanitary landfill is to cover refuse before insect larvae laid on it can hatch; and before putrefaction can proceed far. The cover reduces the problems of blowing paper and plastic; of dump fires; of rat infestation; and of odors. The quantity of cover material required is estimated by Los Angeles' County engineers, who manage probably the largest basic-sanitary-landfill operations in the world, as "one part earth to four parts refuse" (which is the title of a valuable film produced for the Bureau of Solid-Waste Management [2]).

Basic sanitary landfill is suitable for large and small ton-per-day flows, although there are obvious savings to be given by large operations. Figure 6.1, taken from Sorg and Hickman [3], shows the order of costs.

Sufficient cover material must be available in a suitable form, which means that basic sanitary landfill poses some difficulties in areas where severe winters cause stockpiles to become frozen hard.

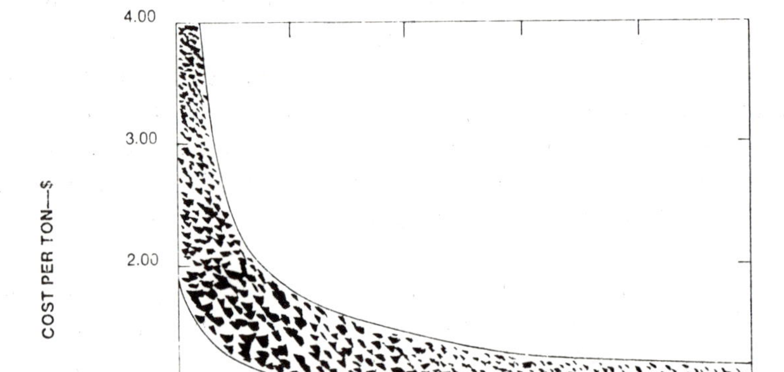

¹ Based on 6-day work week.
² Based on national average of 4.5 lbs per person per calendar day.

Figure 6.1 Basic sanitary-landfill operating costs.

courtesy Bureau of Solid-Waste Management/HEW [4]

Baling and Landfill

Baling is an apparently minor variation of basic sanitary landfill, and is used in connection with some types of transfer-station designs applicable especially when there has to be a long haul to the landfill. At the landfill site, compaction vehicles are not required except to a minor degree for the covering (and crack-filling) material. A mechanical-handling device such as a fork lift or front-end loader may be needed.

Few landfills have operated with bales only, and most of those that have, have been on an experimental basis – for instance, in San Diego [4]. Accordingly, it is difficult to draw firm conclusions as to the merit of baling and as to the optimum baling conditions. However, it is safe to say that bale landfills should be good neighbors, since blowing paper, insects, rats, and fires should all be even less of a problem than for basic sanitary landfills. The even surfaces of well-compacted bales may lead to a lower cover-material requirement, but much depends on how well the bales are fitted together. Decomposition should be very slow, which is presumably good from the point of view of explosion and poisoning hazards resulting from the buildup of methane concentrations. Because highly compacted bales are relatively impervious to water, leachate should be of low concentration. Where it is planned to build on a completed landfill, the uncertain rate of settling of bale landfills may present problems, or alternatively may require a form of construction which bears on the substrata.

The baling transfer station being brought into operation at the time of writing by Reclamation Systems, Inc. [5] in Cambridge, Massachusetts, uses two vertical Lombard presses with a total capacity of 2,000-3,000 tons per day, and has cost a total of \$2.5-\$3 million, (Figure 6.2). This is less than ten percent of the first cost of an equivalent incinerator. Initial charges are between \$8 and \$10 per ton for baling and disposal. Reclamation of some paper and steel is practiced.

Baling has been investigated extensively in connection with long-distance railhaul of solid wastes by several groups. The most detailed study is by Wolf, for the American Public Works Association and the U.S. Public Health Service, [6].

Figure 6.2 R.S.I. compaction system.

Landfill with Pulverized Refuse

Refuse can be comminuted (reduced to pieces smaller than in the original, also known as pulverized, milled, shredded or ground) to a wide range of mean particle sizes, and then landfilled. This type of landfill is gaining popularity in Europe and has been successfully experimented with in this country; particle sizes of generally one to six inches are used. The comminution normally takes place in a transfer station where the collection trucks deliver mixed refuse. The most common form of grinder or pulverizer is a hammer mill (Figure 2.10), fed by conveyor belt and using another conveyor belt to deliver the reduced refuse to a delivery vehicle. Figure 6.3 shows the Heil-Tollmache installation at Madison, Wisconsin.

Comminution has been found to give the following advantages when used for landfill.[1]

1. A substantial degree of compaction is achieved (the uncompacted refuse has a density similar to the compacted refuse as delivered by "packer" trucks; a small compaction pressure can then double the density giving, for instance, 1,000 lbm per cu. yd. for a pressure of 50 lbf per sq. in. [4].
2. Insects do not lay eggs, or, if they do, the larvae do not hatch in large numbers [8].
3. Rats and other vermin cannot burrow and nest in milled refuse, apparently because holes collapse and because the presence of broken glass is a deterrent [8]. (Figure 6.4a)
4. Underground fires (even when experimentally set) are smothered and extinguished [9].
5. There is almost no blowing of paper and dust.
6. Rapid breakdown of the organic constituents takes place, which leads to rapid settling, and to an early capability of the filled land to be used for a variety of purposes.

Figure 6.3 Heil Tollmache hammer mill in Madison, Wis.

courtesy Heil Co.

[1] It has been claimed that no daily cover – only final cover – is required for milled refuse. However, in a recent statement of policy [22] the Environmental Protection Agency called for daily cover, as for basic sanitary landfill, until conclusive data show that this practice is unnecessary.

7. Milled refuse is better than "raw" refuse as a surface on which heavy vehicles can drive before dumping – even in wet weather there is little tendency for the vehicles to bog down, and there is virtually no tire or suspension damage (Figure 6.4b).

Figure 6.4a Duck-raising brood on Madison landfill testifies to absence of rats.

courtesy Heil Co.

Figure 6.4b Heavy vehicles do not sink in pulverized landfill.

courtesy Heil Co.

These advantages have been claimed for European trials of pulverization and have been confirmed in the Madison experiments in the U.S. and in the commercial operation at Montreal [10]. The advantages are obtained at a certain cost, which is that of the transfer-station operation, including the capital and operating cost of the pulverizers. A correlation of the various costs cited in the above references together with a cost breakdown by Ohlgren [11] gives the overall costs shown in Figure 6.5.

The principal reasons for the claimed advantages for the comminution of refuse are that the voids in boxes, cans and bottles are eliminated by size reduction, so that the refuse collapses to a smaller volume. A larger surface area is exposed so that aerobic bacteria, working much faster

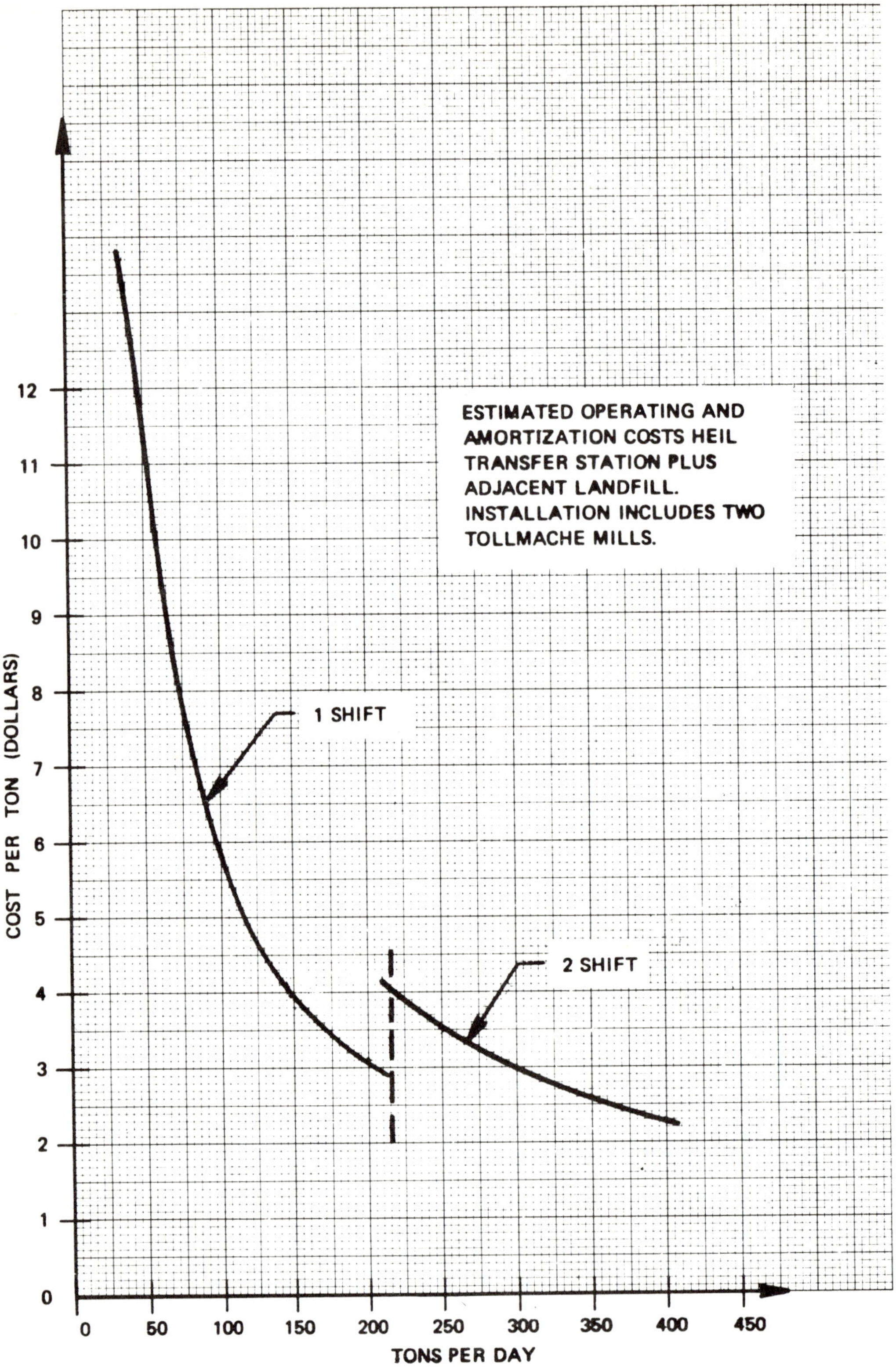

Figure 6.5 Cost estimate for a milling transfer station.

courtesy S. D. Senturia

than their anaerobic counterparts, can produce initial decomposition before the depth of filling has increased to the point where oxygen is excluded. (One would therefore expect that rate of filling would have a strong effect on the settling rate, but no reliable information of this is known at the time of writing).

Degrees of pulverization other than that required to give a one-inch mean particle size can be used. The Milpac hammermill-compactor made by the SFM Corporation produces a particle size of the order of one-eighth inch and compacts the refuse within the machine to solid briquettes, (Figure 2.11). This machine is designed to be used in apartment houses, hotels and hospitals, for instance, to reduce storage and handling costs of solid wastes, but the material produced should be advantageous in land filling. It is unlikely, however, that a situation would arise where a landfill was supplied completely by machines producing a particle size as small as one-eighth inch, so that any special characteristics of milling to this degree of fineness is probably irrelevant.

At the other end of the scale, some continuous-action packer collection trucks achieve a certain degree of size reduction through the continuous tearing action of their compaction screws. Two European examples are the Kuka Shark, which has a continuously rotating drum carrying screw-thread-type projections (Figure 6.6) and the Norba, which has a central meat--grinder type of screw. Were a landfill to be wholly supplied by vehicles of this type, it would be of much interest to see whether or not the landfill displayed the characteristics of one made with milled refuse.

Figure 6.6 Kuka shark shredding collection vehicle.

courtesy St. Regis Paper Co.

Multiple Use of Land

No suggestion made in the M.I.T. summer study on solid wastes [12] created more controversy than that made by Clarkeson regarding the possibility of using rights of way for waste disposal in certain circumstances. However, no valid objections to his suggestions have so far been sustained. While his suggestions are not strictly a method of landfill, but rather suggest potential sites, they deserve mention here.

The types of rights-of-way land which seemed to merit attention were the borders of state and interstate highways (and associated land such as that surrounding interchanges); and the reserved land under or over high-voltage transmission lines, and gas and water pipelines. Along such rights of way there are always areas, sometimes substantial areas, where closely controlled filling would be not only beneficial but quite practicable. For instance, in undulating or hilly country roadside borders are frequently well below the level of the shoulders of the road itself. A hazard is thereby created, because a car veering off the roadway for some reason, such as a front-wheel blowout, might roll over on its way down the embankment and severely injure or kill the occupants. Where such areas can be reached from a suitable approach road (not the highway itself, obviously) it would be wholly beneficial (assuming the absence of other negative factors such as the need to avoid water pollution by leaching) to fill the area, preferably with bales or pulverized refuse, recover with topsoil and replant. Clarkeson estimated that there were sufficient suitable sites along state highways within a 15-mile radius of Boston to accept the whole of the Boston-area refuse for five years.

Pulverized refuse has been suggested as being ideal for this application because of the virtually guaranteed absence of blowing paper and plastic, of fires and dust, and of birds and rats, all of which would be considerable nuisances and possibly would be dangerous if conventional "basic" sanitary landfill were practiced near highways. Baled refuse would presumably also be effective.

EXPERIMENTAL METHODS

The above methods are generally available, while those to be described below have been applied either on a purely experimental basis or are in use in only one location at the time of writing.

Continuous Burial

Two special-purpose machines have been developed for depositing refuse as received from collection trucks under the surface of the ground. In the D&J Press machine, refuse is continuously placed under the surface of open land, and the top soil is replaced as the machine advances, [13], Figure 6.7. The amount of refuse put into the land is limited to a depth of

Figure 6.7 Continuous refuse-burial machine.

courtesy D & J Press

about three feet in the compacted state. In certain circumstances, this may be all that is permissible, and use of the machine would reduce the cost involved in site preparation and in collecting, delivering and placing cover materials.

The second machine in this category is known as the Mole [14] and is in use in King County, Washington. The machine operates in a trench 15-feet deep which is first excavated at the landfill site by a backhoe, Figure 6.8. The refuse is squeezed in a press and is extruded into the trench in nine-feet-square lifts. The Mole is an automated version of the "deep-hole" method of landfill, as used in Lafayette, Louisiana, where a pit 40-feet deep is first excavated by a dragline. The area required for landfill has been claimed to be reduced by 75 percent by these methods. Obviously they will be suitable only in some site conditions. For instance, there will be little possibility of controlling water pollution, although the amount of leachate should be relatively small.

Figure 6.8 "The Mole" compaction-burial machine.

courtesy Hayes Evans, King County, Washington

Pulverization in the Land

A method being investigated by Hart [15] is to apply comminuted refuse at low loadings (up to 400 tons per acre) and to till it into the topsoil. The soil bacteria and the available oxygen produce rapid decomposition of the organic refuse, which then acts as a soil component. At this point, which has been found typically to occur within the range of nine to fifteen months, a new load of pulverized refuse can be added and tilled in. Some of the inorganic components rust or otherwise degrade; the glass, if milled finely, is similar to sand, while most of the plastics and aluminum alloys remain little changed.

A machine which might prove appropriate to this method of waste management is the Roto-shredder, Figure 6.9 developed initially for comminuting material to be composted by the windrow process that further developed for further municipal refuse. It is essentially a traveling grinder which moves over and through the pile of refuse left by a conventional packer truck after it has unloaded and, in one or two passes, reduces the refuse to pieces covering a range of sizes depending on the materials. It can be used for direct pulverization at the landfill, but in application to refuse stabilization in the land it would seem particularly appropriate.

Figure 6.9 Roto-shredder travelling refuse pulverizer.
courtesy Cobey Environmental Control, Inc.

The roto-shredder operating at the Crawford county landfill (Ohio), shredding and composting unsorted city refuse on a Unite States public health service grant.

Rapid Stabilization by Aeration

Stone [16] and the City of Santa Clara, California, carried out a three-year trial of a system involving mechanical aeration of municipal and commercial refuse. Two aeration cells, approximately 65-feet wide by 75-feet long and 17-feet high were constructed having an air-distribution system consisting of a grid work of four-inch-diameter perforated clay pipes placed in gravel-filled trenches over the base of the cells. A five-horsepower blower was used for air supply and the raw refuse was deposited in the cell by regular municipal collection trucks. When one cell was full, it was covered with one foot of soil and filling proceeded in the second cell. Within a few days the temperature rose to between 140F and 200F; both the temperature and the gas composition were monitored during the tests. After 90 days, it was determined that the majority of the decomposition had occurred and the contents of each cell were bulldozed over to an adjacent fill area.

Stone's conclusions included the following.

1. The incremental costs of aeration over basic sanitary landfill were 0.53 dollars per ton for 1966-69 Santa Clara operating and construction costs. This figure assumes no credit for the stabilization and volume reduction achieved.
2. The volume reduction of 20 to 40 percent occurred for "normal" municipal rubbish when aerated for 90 days; no significant volume reduction was found with "commercial" refuse.
3. Soil cover was necessary during aeration to control fires.

4. Vermin, flies and other health hazards were controlled by high temperatures and oxidation of the putrescible solid-waste components. Pathogenic bacteria were killed by the high temperatures.
5. The odors given off were predominantly sweet and tobacco-like.

In addition to the credits for decreased refuse volume and for the greater availability of the filled land for high-value use due to the decreased amount of settling to be expected, there should additionally be evaluated the decreased water pollution that would result from the low BOD of the leachates, or alternatively the low cost of providing leachate treatment. No information on leachates was reported.

PROBLEMS AND PRECAUTIONS IN LAND FILLING

Water Pollution

Research has shown that sanitary landfills will invariably produce a leachate. This leachate may be 100 times as strong as sewage, with an initial BOD of over 20,000 milligrams per liter, [7]. Although the BOD has been found to be reduced by 95 percent after a percolation of 12 feet through soil, this oxygen demand is still higher than that of raw sewage; in addition, there is no certainty that the leachate may not find its way directly to some underground water course which might communicate with an acquifer. Accordingly, a number of precautions should be taken when selecting and preparing a site for a sanitary landfill.

1. The lowest part of the landfill should be several feet, of the order of 10-feet, above the highest local water-table level.
2. If the walls of the landfill are not all naturally impervious, they should be lined, perhaps with clay, and sand, gravel, or tile drains should be laid to draw off leachate for ducting to a sewer or for other treatment.
3. Surface drainage from adjacent tributary areas should be diverted from the site or carried in a lined channel or storm drain to prevent storm waters from entering the rubbish, [18].

Emrich and Landon, (op cit) suggested a fourth possibility, which is to site the landfill above an area where there is a large volume of subsurface water movement so as to dilute the leachate to safe levels. The reliability of this procedure would seem to be open to some doubt, however.

Studies and experiments carried out at Drexel have shown that leachate concentration varies greatly for any one landfill and may still be significant after 20 years, [19].

Gas Production

All landfills will produce some gas, primarily methane, which may travel horizontally for considerable distances if it encounters a suitable seam in the surrounding strata and may gather in pockets or in building basements, for instance, to form a poisoning and explosion hazard. These dangers are, of course, much greater when building takes place on a completed landfill. The following recommendations are taken from [18].

1. Comprehensive soils information should be obtained for the design of new sanitary landfills to evaluate the surrounding conditions and to anticipate the movement of decomposition gases to adjacent areas. If significant gas movement is expected, consideration should be given to the design of a soil barrier to prevent movement of gases to adjoining property.
2. A program of gas monitoring should be undertaken at all completed landfills regardless of

age since completion. This monitoring could be performed using methane-detection equipment of the type used by sewer-maintenance agencies, fire departments, or gas-distribution utilities. Results of this program should be brought to the attention of building officials so that steps can be taken to prevent construction of buildings without proper precautions.

Methane-gas production and movement can be detected by "sniffing bar holes" in the upper layers of soils on or adjacent to completed landfills. Sewers and storm drains in the vicinity should also be checked for the presence of gas. Even if gas is not detected, the monitoring should be repeated periodically and the site should be watched for signs of decomposition. If, however, gases are found, more detailed study of the landfill in surrounding areas should be made to determine the extent of gas movement and to evaluate the hazards, if any.

3. Where sanitary landfills are in a known state of decomposition, and where there are potential hazards, steps should be taken to provide ventilation beneath and in all structures in the affected zone to prevent the accumulation of explosive mixtures, odors, and possible fires.

Fires

In a "normal" landfill, a fire, once it has taken hold, may smolder on for many years, occasionally becoming extensive and sometimes apparently dying down. Fire officials have found no certain way of dealing with such problems short of massive digging. They usually have to content themselves with applying large quantities of water where the smoke and flames appear, which may be quite a distance from the center of the conflagration. The cellular structure of a "basic" sanitary landfill prevents a fire traveling very far, and the sealing of the top of the cell may be enough to starve a fire sufficiently to put it out. Bale or pulverized landfills are also resistant to burning.

Settling

Settling is to be expected on all types of landfill except land composed of highly compacted bales, the settling or expansion behavior of which has not yet been determined. Even for conventional dumps, there appear to be no reliable guidelines as to how much or over what period settling might be expected. "Wet" quarryholes in Quincy, Massachusetts have sunk 15 feet in a year after being filled to a depth of 125 feet over a period of about two years. Similar quarryholes which have been first pumped dry and filled on a "sanitary" schedule have shown only five to ten feet of settling [20]. Elsewhere landfills have been excavated for construction and sometimes virtually no decomposition of newspapers which had been in place for 40 years has been found. Therefore, while aeration or pulverization or both may be used to accelerate decomposition and settling, for existing landfills there seems to be no good way of predicting a pattern of settling. Special construction methods which do not rely on the properties or the location of the upper layers have to be used for buildings constructed on landfills.

Miscellaneous Problems

An earth "berm" (bank), and/or a fixed shield of bushes and trees, and/or a high fence, are solutions for a number of additional problems that can arise with landfills [21]. Such shields will cut down the problem of blowing paper and dust; will attenuate the noise from earth-moving and compacting machinery and from the refuse trucks which might operate on a landfill; and will reduce the unsightliness of dumping operations. Unauthorized entry for salvage or play will also be deterred. The community benefit that results from individual efforts at recycling is completely overbalanced by the very high risk of injury and death experienced by people scavenging (or playing) on dumps.

Some well-managed and well-laid out landfills include a curved access road so that the line of sight and sound prevents dump operations being seen and reduces the noise radiated directly [23].

The offensive smell of some dumps may result from smoldering fires, from blowing dust, or from anaerobic composition, all of which are controlled by adequate cover being used.

CONCLUSIONS

Landfill is the most prevalent and still in many ways the best method of solid-waste disposal. It can be brought into higher public favor by a combination of good management and modern techniques. Many new approaches can be tried. If landfills in all areas could be brought up to the standard of, for instance, those of Los Angeles County, citizen resistance to landfills would largely disappear and large areas would be made available for landfilling.

REFERENCES

1. U.S. Health Service. "Final Report of Board to Investigate the Landfills Maintained in the Borough of Queens." Washington, D.C., May 1940.
2. Los Angeles County Sanitation District for the U.S. Environmental Protection Agency. "One Part Earth to Four Parts Refuse." *In* film with narrative; Rockville, Maryland, 1966.
3. Thomas J. Sorg, and H. Lanier Hickman, Jr. "Sanitary-Landfill Facts." U.S. Public Health Service, 1970.
4. City of San Diego and U.S. Public-Health Service. "Baling Municipal Refuse." Report under PHS contract no. 1-DO1-U1-00061-01, April, 1968.
5. Joseph LaBarbara "Solid-Waste Compaction." *In* New Directions in Solid-Waste Processing, pp 152-162, University of Massachusetts, May, 1970.
6. Karl W. Wolf, "Rail Transport of Solid Wastes." Interim report, phase 1; American Public-Works Association for the U.S. Health Service, 1969.
7. P. K. Patrick, "Waste-Volume Reduction by Pulverization, Crushing, and Shearing." Proceedings; The Institute of Public Cleansing 69th Annual Conference. Blackpool, England, June, 1967.
8. R. K. Ham, "The Relative Attractiveness of Milled and Non-Milled Refuse to Rats and Flies." *In* Public Works, pp 74-5, July, 1969.
9. John J. Reinhardt, and Gerard Rohlich. "Solid-Waste Reduction/Salvage Plant – an interim report." U.S. Public-Health Service, 1968.
10. W. J. Johnson, "Refuse-Reduction Plant, Montreal-Quebec." *Engineering Journal,* June, 1969.
11. Personal communication. Ray Ohlgren, The Heil Company, August, 1970.
12. David Gordon Wilson, Editor. "Summer Study on the Management of Solid Wastes." Urban Systems Laboratory, M.I.T., September, 1968.
13. C. H. Billings, "Operation 'Big-Squeeze' takes on Refuse Disposal." *In* Public Works v. 97, no. 87. January, 1966.
14. H. Evans, Jr. "A New Idea in Landfill Operations." *In* The American City, v. 82, no. 114. March, 1967.
15. S. A. Hart, "Refuse Stabilization in the Land." American Society of Mechanical Engineers. paper 69-WA/PID-5, 1969.
16. Ralph Stone, Samuel N. Cristofano, and E. T. Conrad. "Mechanical Aeration of a Landfill." American Society of Mechanical Engineers. paper 69-WA/PID-20, 1969.
17. Grover J. Emrich, and Ronald A. Landon. "Generation of Leachate from Landfills and its Subsurface Movement." *In* Proceedings; Annual Northeastern Regional Antipollution Conference, July 22-25, 1969. pp. 57-63.

18. Los Angeles County and Engineering Science, Inc. for the U.S. Public-Health Service. "Development of Construction and Use Criteria for Sanitary Landfills – an Interim Report." Arcadia, California, 1969.
19. R. L. Stanier, A. A. Fungolari, R. I. Schoenberger, and P. W. Purdom. "Criteria for Sanitary-Landfill Development." *In* Public Works. March, 1971. pp. 77-79.
20. Jack Brown, public works director, City of Quincy, Massachusetts, Personal communication. March, 1971.
21. "Kenilworth Model Sanitary Landfill." Interim Report, U.S. Public-Health Service, 1969.
22. Richard D. Vaughan, "Statement of Policy – daily cover for milled solid wastes." Environmental-protection Agency, Solid-wastes Office, February 2, 1971.

CHAPTER 7

MUNICIPAL INCINERATION

Incineration as one of the older methods of solid-waste disposal has had more than its share of abuse and exaggerated claims. It is at times blamed for most of the air pollution in cities and at other times proposed as a SO_2-pollution-free source of heat and electricity. Wide-ranging claims are also made concerning costs, with initial capital costs cited at anywhere from $3,000 to $20,000 per ton per 24-hour-day (TPD) installed capacity, and disposal costs cited at values ranging from a small positive credit (for some proposed designs with heat recovery) to $15 per ton.

There is some basis for the widely varying claims. Many incinerators are poorly designed, poorly operated and without air-pollution control. Burnout in such units may be so poor that newspapers and telephone books are still legible in the residue. In contrast, modern well-managed units are equipped with efficient flue-gas scrubbers or, in a few plants, electrostatic precipitators for air-pollution control. These units sometimes recover energy from the waste gases and produce a sterile ash which can be used for ballast or fill.

In this chapter, some of the factors of design, operation and financing which influence the overall performance of incinerators will be discussed.

THE ROLE OF INCINERATION

Incineration's primary functions are to reduce volume and mass and to sterilize. Approximately 20 percent by weight of refuse remains as residue when complete burnout is achieved. The average density of refuse is 13 to 17 lbm per cu ft as fired, and the ultimate density of the residue may be as high as 110 lbm per cu ft for an incinerator in which the residue is fused, as a consequence of which the maximum volume reduction, due to both weight decrease and density increase, is about 97.5 percent. For conventional, low-temperature, incinerators, the volume reduction is not as high, but still is in the range of 80 to 95 percent.

REFUSE AS A FUEL

The composition of a typical mixed domestic and commercial refuse is presented in Table 7.1 both by category and component.[1] The characteristic properties of refuse are compared with those of some selected fuels in Table 7.2. Relative to coal, refuse is a low-sulfur, low-density, low-heating-value, high-ash-content, high-moisture-content fuel. The heating value of refuse on a volume basis is much closer to that of peat than of coal. Probably the largest disadvantages of refuse as a potential energy source are its heterogeneity, variability,[2] and small supply (relative to energy demand). Most fossil-fuel boilers operate with reliability factors which are hard to achieve with a material as difficult to handle and as variable as refuse.

---

[1] For details on the composition of refuse, its variation with season, region, and weather, see [1, 2 and 3].

[2] Charges to a furnace may vary from wet leaves, with a negative net heating value, to plastics or gasoline-soaked fabrics, with heating values of 19,000 Btu/lbm and burning rates high enough to be the cause of a mild explosion.

Furthermore, the total potential energy supply from refuse could at best supply only three percent of the nation's energy requirements. (The total U.S. energy consumption is greater than the energy equivalent of two-billion tons of coal per year; the total commercial and domestic waste generated has an energy equivalent of about 60 million tons of coal per year). The conditions under which energy from refuse incineration can be recovered economically will be discussed in a later section.

TABLE 7.1

REFUSE COMPOSITION [Ref. 1]

Category	Weight % (as fired)*	Component	Weight %
Metal	8.7	Moisture (H_2O)	28.16
Paper	44.2	Carbon (C)	25.62
Plastics	1.2		
Leather and rubber	1.7	Oxygen (O)	21.21
Textiles	2.3	Hydrogen (H)	3.45
Wood	2.5	Sulfur (S)	0.10
Food waste	16.6	Nitrogen (N)	0.64
Yard waste	12.6	Ash	20.82
Glass	8.5		
Miscellaneous	1.7		

*This weight distribution shows the effects of moisture transfer between the categories in the refuse during storage and handling. For example, the food waste tends to lose moisture and the paper absorbs moisture.

TABLE 7.2

CHARACTERISTIC PROPERTIES OF REFUSE AND SELECTED FUELS

Fuel	Moisture (%)	Ash %	Density (as fired) (lbm/ft^3)	Lower heating value (Btu/lbm)	Lower heating value (Btu/cu. ft.)
Refuse	28.16	20.82	17.0	3,935	67,000
Wood	46.9	1.5	60.5	3,190	236,000
Peat	64.3	10.0	25.0	2,390	59,800
Bituminous coal (high-volatile)	8.6	8.4	50.0	11,735	585,000
Fuel oil (#6)	1.5	0.08	61.5	17,280	1,060,000
Methane	0	0	0.042	21,555	1,050

DESIGN CRITERIA FOR INCINERATORS

The basic components in an incinerator are, as shown in Figure 7.1, an unloading area, a storage bin, a refuse-feed mechanism, a combustion chamber, a large air-supply system, a

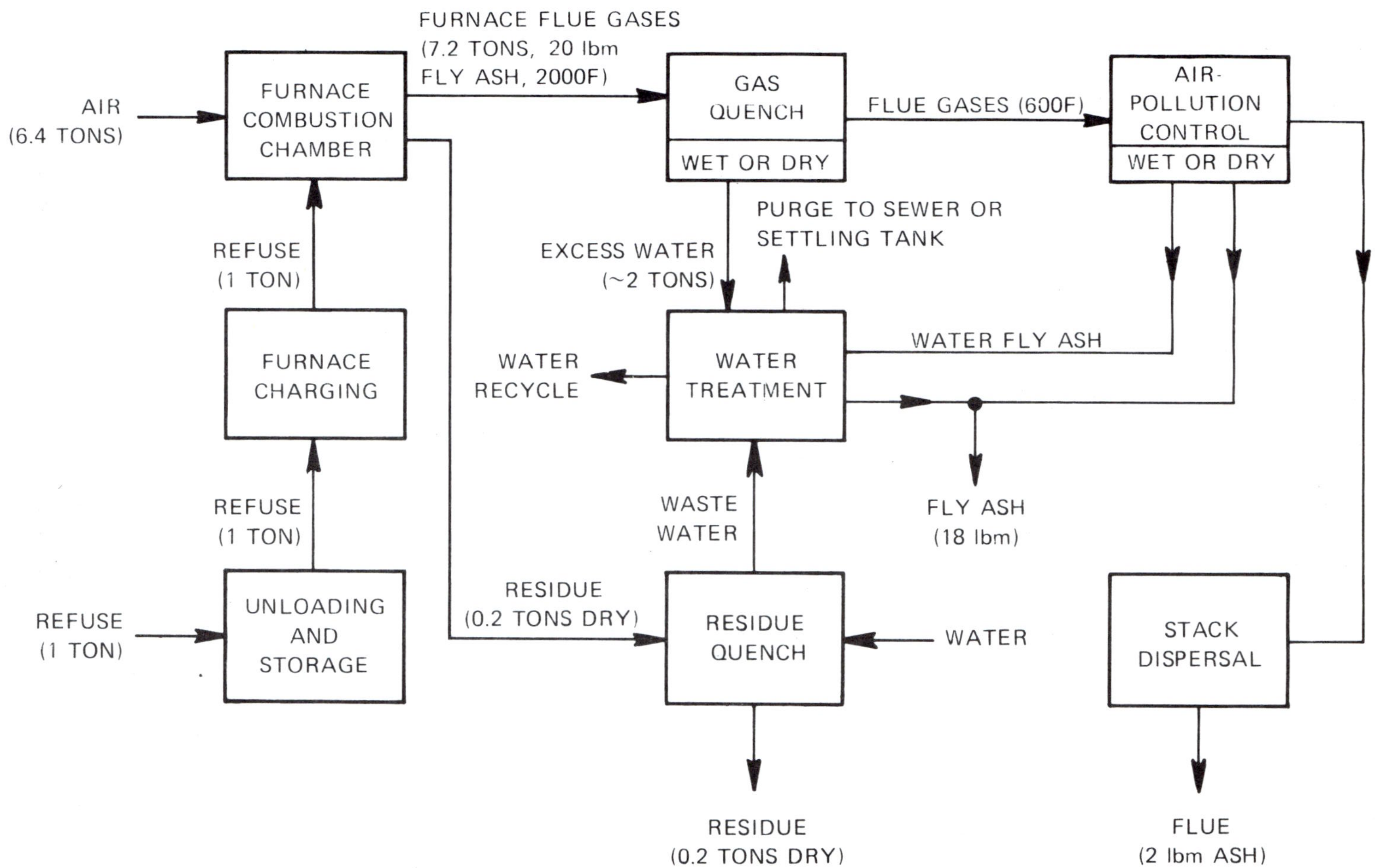

NOTE: Quantities in parentheses are rough measures of flow rates and temperatures.

Figure 7.1 Basic components of an incinerator.

residue-quench and residue-disposal system, and flue-gas- and water-treatment systems. The flue gases may typically contain about 20 lbm of particulates per ton of refuse and must be cleaned before dispersal into the atmosphere. If a wet-scrubbing system is used, the problem of water treatment must be considered since the water will absorb highly corrosive elements, primarily HCl, from the flue gases. For a dry-gas treatment, such as mechanical cyclones or electrostatic precipitation, flue gases must be cooled before treatment to the temperatures, usually in the range of 450F to 700F, that can be tolerated by the air-pollution-control device.

Excess-air Requirements

Some insight on the design constraints on incinerators can be gained from consideration of the values for air requirement and adiabatic flame temperature given in Table 7.3. If refuse could be burned with the minimum amount of air required to complete the combustion (zero-percent excess air), for every ton of refuse burned 3.21 tons of air would have to be supplied, and 4.03 tons of flue gas initially at temperatures close to 3,000F would have to be processed by the gas-quench and air-pollution-control systems before discharge into the atmosphere.

TABLE 7.3

STOICHIOMETRIC RELATIONS AND ADIABATIC FLAME TEMPERATURES FOR THE COMBUSTION OF A TYPICAL REFUSE WITH VARYING AMOUNTS OF EXCESS AIR*

Percent Excess Air	0%	50%	100%	200%	300%
Air requirements (lbm air per lbm refuse)	3.21	4.81	6.42	9.63	12.84
Flue-gas generation (lbm flue gas per lbm refuse)	4.03	5.64	7.25	10.45	13.65
Adiabatic flame temperature (F)	3,020	2,450	1,990	1,460	1,180
Dew point (F)	147.0	135.0	126.0	113.0	104.0

*Refuse Composition assumed for these calculations as in Table 7.1

Because of difficulties in metering air to the fuel at the desired locations in the exact amounts required, most combustion systems operate with excess air in proportions which usually increase with the difficulty of mixing the fuel with the air. The minimum proportions of excess air that have been reported for incinerators are about 50 percent, somewhat larger than the amounts of 30 percent used for lump-coal combustion, 15-20 percent for pulverized-coal combustion or 0-5 percent used for gas and oil combustion.

The adiabatic flame temperature for refuse combustion at 50 percent excess air is 2,450F, a value at which the fly ash carried in the flue gases is in its molten state and can cause severe damage to refractories. Therefore, operation at low excess air is restricted either to furnaces whose combustion chambers are lined with water-cooled tubes or to high-temperature slagging incinerators where special designs of the furnace enclosure are required to avoid or resist the fluxing of the molten ash with, or adherence to, the refractory. Conventional, refractory-walled furnaces operate below the fusion temperature of ash and it is customary to aim for the maximum flue-gas temperature to be 1,800F to 2,000F. The flue gases should in every case be maintained above 1,200F in the combustion chamber in order to destroy any noxious gases. These constraints on temperature indicate that the range of excess air for a refractory-lined furnace could be 100 to 250 percent.

The lower values are preferred because of the economies to be realized by handling smaller gas volumes. On account of the inadequate gas-mixing processes in incinerators, however, excess air levels above 250 percent are often used to avoid local temperature extremes in vulnerable parts of the furnace.

Drying and Burnout Rates

The size of the combustion chamber and the residence time of the refuse are further constrained by the rates of drying, pyrolysis, and burnout of the refuse elements on the grate and by the kinetics of oxidation of the pyrolysis products above the bed. A schematic of the processes occurring in and immediately above the bed is shown in Figure 7.2. A quantitative model of the refuse-burning process is not yet available. Typically, residence times on the grate of 40 to 60 minutes are required, with higher values needed when burning either bulky objects or refuse with a high moisture content. Good mixing is required above the bed in order to complete the burning; it may be provided by overfire jets such as shown in Figure 7.2. Much remains to be learned, however, on how to control the burnout quality by adjustment of the overfire and underfire air rates, the air distribution below the grate, the grate speed, and refuse-feed rate.

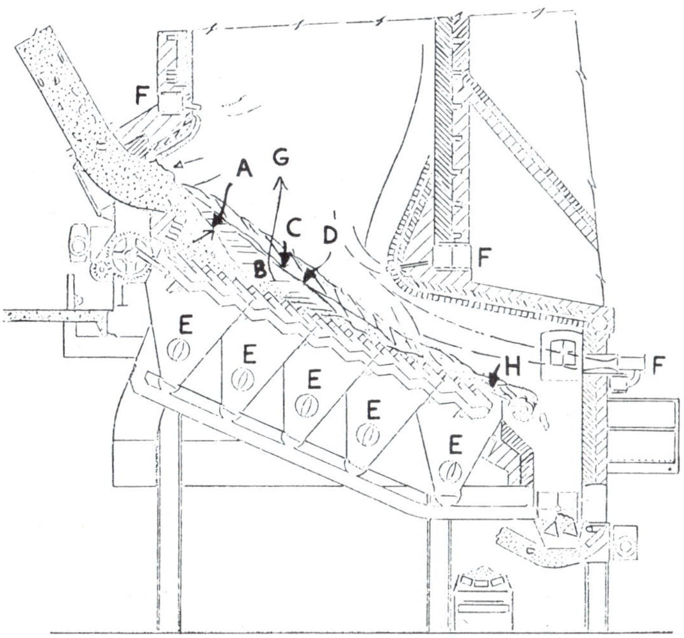

LEGEND

A. Drying wave (irregular due to heterogeneity of refuse and to poor mixing)
B. Pyrolysis zone
C. Ignition wave (irregular)
D. Burnout zone
E. Wind boxes for under-fire air
F. Over-fire air nozzles
G. Pyrolysis products burning with over-fire air
H. Burned-out residue or ash

Figure 7.2 Representation of processes occurring on incinerator grates [24].

Severity of the Design Problem

It becomes readily apparent that, although the burning of refuse is seemingly so easy to accomplish in a backyard incinerator where no consideration is given to air pollution, residue quality, or labor requirement, the efficient disposal of solid wastes at the municipal scale requires the design of fairly elaborate solid-handling, gas-handling, and possible water-handling systems. The number of design configurations comprising the existing plant inventory is large, partly because every incinerator has been tailored to local conditions and biases and partly because the poor understanding of the basic design parameters has not permitted an unequivocal system selection. The following section will present a very brief description of the options available in order to define some of the terms and to develop an awareness of the difficulty in decision-making by both design engineer and municipality. More comprehensive coverage of the material can be found in [1, 4, 5, 6, and 7].

INCINERATOR COMPONENTS

Superstructure

The location of incinerators has varied from residential areas to landfill sites and has resulted in varying demands for aesthetically pleasing confining buildings. At times, efforts to overcome neighborhood objections have resulted in a disproportionate fraction of the initial cost being channelled into a handsome edifice at the expense of the functional components of the incinerator. At the other extreme, for units in isolated or industrial sites, the confining building can be very simple or largely eliminated. From a viewpoint of heat rejection from the hot combustion products, a furnace in the open with thin refractory walls losing energy directly to the environment may be optimal [8].

Receiving and Storage Area

The storage pits for holding the refuse until charged into furnaces should have sufficient capacity to maintain the fires between collection periods. Recognition of the economy of continuous operation and of reduced refractory wear when temperature cycling of the furnaces is eliminated has resulted in an increased trend to 24-hour and in some cases seven-day-per-week operation. Storage bins for continuous operation must therefore provide sufficient capacity to supply the furnaces over a weekend, or about 250 cubic feet per TPD operating capacity.

The storage volume is usually provided by a pit located between the furnaces and a tipping area for the trucks. A traveling crane and bucket are used to charge the furnace. Sometimes, at the discretion and according to the motivation of the operator, the crane and bucket are also used to mix the refuse coarsely within the pits.

There are many refinements that have been incorporated into the design of the receiving and storage area. To eliminate possible nuisance to the neighborhood, and to provide protection from the weather, the tipping area is frequently covered; in a few installations dust and odor are controlled by maintaining the storage pits under reduced pressure. This can be achieved by discharging the compactor trucks into the pit through chutes which are sealed when not in use and by drawing the furnace air from a position high in the pit. Regular cleaning and maintenance of the ductwork, fans, grills and so forth is required when this technique is used, however, as a consequence of the high dust loading of the air drawn from this area.

Some smaller incinerators use a paved floor instead of a pit to store the refuse. In such units, the refuse can be pushed onto a conveyor for loading into the feed hopper or pushed through an opening in the floor into the feed hopper.

The large volume required for storage, the high initial cost and maintenance on cranes, and the high salaries of crane operators make the storage and handling of refuse one of the major costs of incineration. Clearly there is room here for the invention of a better system. A belt-conveyor system tried in Birmingham, England [9] met with problems of bridging of the refuse. A rotating-bucket system developed in Germany [10] does not appear to have the desired storage capacity. A bin system such as that developed by Atlas [11] for handling wood and bagasse may be adaptable to refuse feeding in large plants, although pre-shredding, which adds to the cost, is probably required.

Refuse-feed System

The feeding of refuse may be either batch or continuous. Figure 7.3, taken from a recent survey of municipal incineration [1], shows the practically complete displacement of batch by continuous units in new construction over the period of 1955 to 1969. Each square below and above the zero ordinate represents a plant installed in the year given by the abscissa which used batch and continuous furnaces, respectively.

Continuous feeding is usually achieved by a hopper and a refractory-lined or water-cooled gravity chute. The refuse in the chute and hopper provides an air seal. In some batch units, charges are fed into a hopper by a crane and discharged periodically into a furnace by the opening of a charging gate. This batch approach to operation is disadvantageous as the opening of the charging gate usually is accompanied by a large influx of air which cools the combustion products excessively and leads to temporary high rates of emission of combustibles. This problem is compounded by the high rate of evolution of pyrolysis products when a fresh charge is exposed to the hot furnace walls. Also, the cool air chills the refractory thus increasing maintenance. Another major problem arising from the unsteady flue-gas flow rate relates to the selection and control of air-pollution-control equipment. For small installations, however, there may be no practical alternative to batch feeding. In such cases, air-lock seals on the feed system as by a ram through an opening in the side of the furnace and/or the introduction of smaller quantities of refuse at each charge (to stay within the operating limits of the air-supply system) are recommended.

Grate and Air Supply

The functions to be performed in a grate are best discussed with reference to Figure 7.2. The refuse fed to the furnace is first dried and preheated by the radiation from the hot combustion gases and refractory. The undergrate air assists in the drying process partly by its moisture-carrying capacity but, more importantly, by supporting exothermic combustion reactions within the bed. The raw refuse, as it is dried and heated further, pyrolyzes[3] and then ignites. Combustion takes place both on the grate to burn out the residue and in the gas space to burn out the pyrolysis products. Overfire air jets greatly assist the mixing and combustion in the overhead combustion space.

It is apparent that a grate must perform a number of functions: provide support for the

---

[3]Pyrolysis refers to the thermal decomposition of the refuse accompanied by the evolution of combustible gases, to be described in greater detail later.

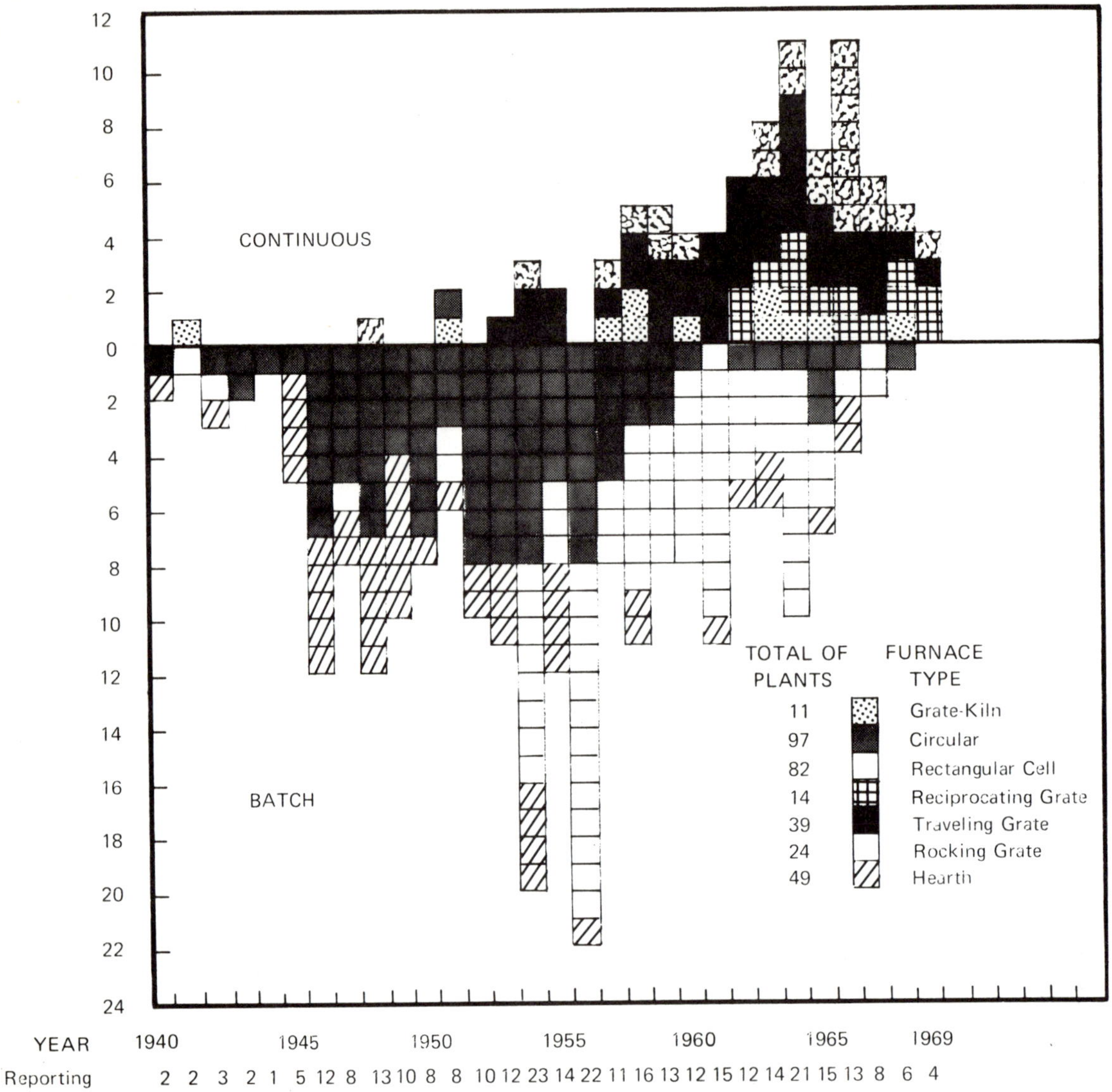

Figure 7.3 Stoker type and furnace feed.

refuse; admit the underfire air through openings; transport the refuse from feed chute to ash quench; agitate the bed to break up clumps; and serve to re-distribute the burning mass. Figure 7.4 shows schematic arrangements of a number of grate designs that are in use for refuse burning. The trend in American practice may be seen from the histogram in Figure 7.3 indicating the number of units of various designs installed each year over the period 1940 to 1969. The tendency has been towards the use of stokers that provide high agitation even though the agitation increases the particulate loading from the furnace chamber (but not necessarily the stack-gas loading since the air-pollution-control (APC) system can be designed to handle whatever loading is encountered in the furnace gases). Agitation is valuable both in

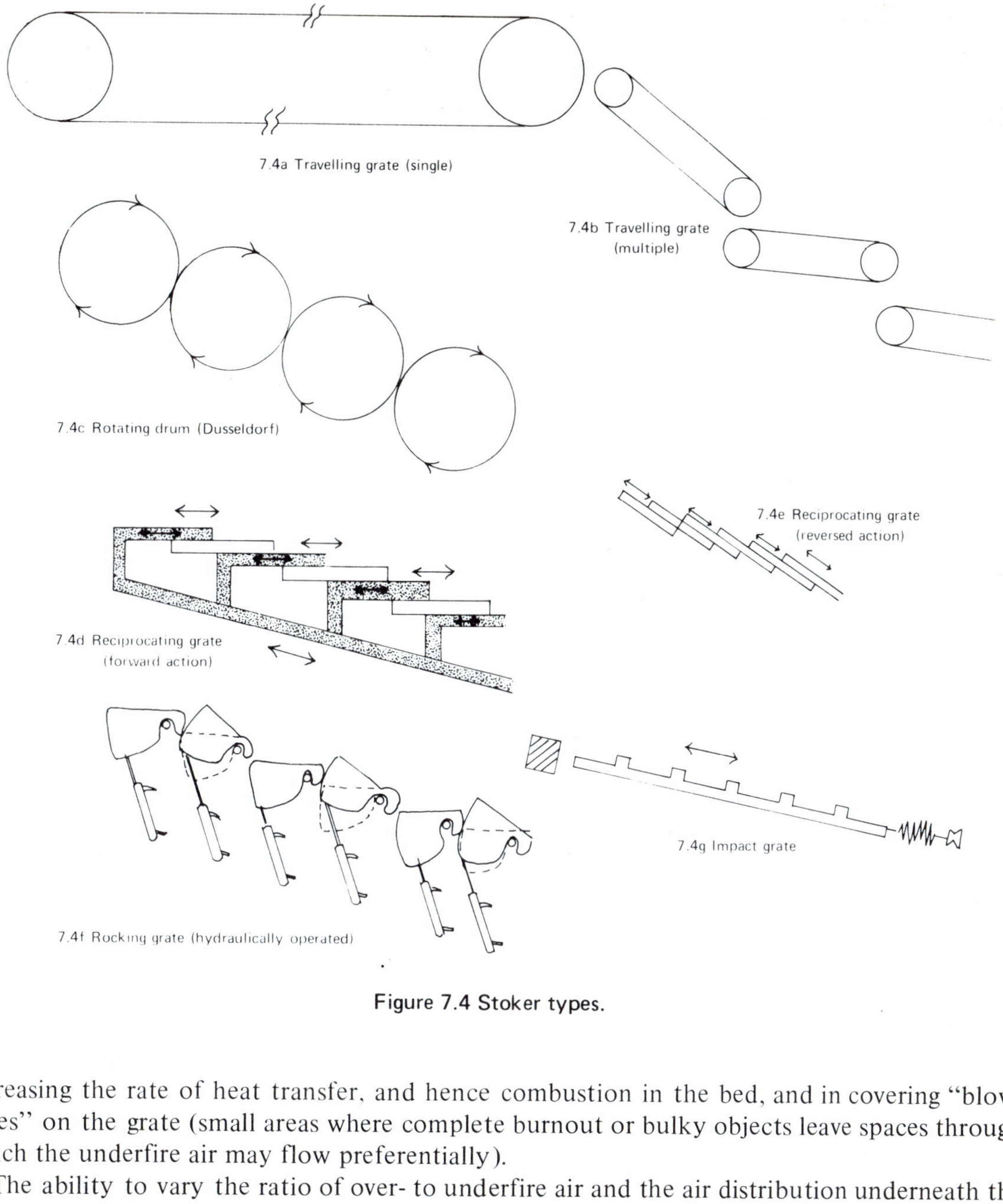

Figure 7.4 Stoker types.

increasing the rate of heat transfer, and hence combustion in the bed, and in covering "blow-holes" on the grate (small areas where complete burnout or bulky objects leave spaces through which the underfire air may flow preferentially).

The ability to vary the ratio of over- to underfire air and the air distribution underneath the grate provides the operator with a flexibility of operation that is rarely exploited. The recognition [12, 13 and 14] that increased undergrate air velocities results in increased particulate entrainment has been used to advocate reduction in underfire air rates. But reduced underfire air rates decrease burning rates (requiring down-rating of the capacity of existing plants or comparative under-rating of new plants) and may lead to hot spots on the grate (with consequence warpage or burnout). The underfire air rate settings must be based on consideration of these factors and the effect of particulate loading on the APC system. The balance of the air is fed through overfire jets. Although the role of overfire air jets for promoting mixing above the bed and for reducing smoke emission has long been recognized in application to coal stokers, many incinerators either do not use high-momentum jets or place them at a position far enough from the grate that the air, far from completing the combustion, actually quenches the incompletely burned gases. There is hope that improvements may be forthcoming in this area.

The burning rates that are conventionally attained in incinerators range from 50 to 80 lbm per hour per sq ft or, for comparison with other solid fuels, 200,000 to 400,000 Btu per hr per sq ft. These rates are surprisingly close to values of 600,000 Btu per hr per sq ft normally encountered in coal-burning stokers.

Combustion Chamber

The combustion chamber serves to confine the flames and to augment the radiative energy transfer to the fuel bed. Refractory walls are most commonly used in the U.S. Increasing consideration is being given, however, to the use of water-cooled walls, since these eliminate the temperature cycling, and consequent wear, on refractories, assist in the control of slagging or fluxing problems, and reduce the volume of gas to be treated by the APC system (see section on gas-quench system) by allowing operation at lower excess air levels. Fife and Boyer [15] have concluded that the savings on refractory maintenance, on quench water, and on the volume of gas to be treated more than offsets the higher initial capital investment for water-wall incinerators when unit capacities exceed 300 TPD; but good comparative statistics on operating units are lacking. Corrosion may, however, be a serious problem with the water-wall units. European experience [17 and 18] has been that tube surfaces both near the grate and in the superheater platens are frequently corroded severely. Tubes have had to be replaced after as little as 1,000 hours of service. The principal causes for corrosion in incinerator boilers is thought to be the HCl evolved by the chlorinated plastics, complex compounds of the alkali metals and volatile oxides and sulfates of heavy metals such as zinc, tin and lead. The extent of corrosion is dependent on tube-metal temperature and on whether the conditions in the furnace gases are reducing or oxidizing, but the mechanism is still not understood. A better understanding of the chemistry of the tube-metal reaction with the fly-ash slag may lead to a selection of alloys resistant to attack. In the interim period protection is obtained by coating the tubes or covering them with a thin refractory layer, by the use of cast-iron blocks over the tubes, or in some instances, by modifying the flow pattern of the gases over the tubes.

Residue Handling

The residue from the grate is usually quenched in water and then either dumped directly into residue-handling trucks or conveyed to an ash storage bin from which it is dumped into the trucks. In a number of installations, in particular in Europe, metal is salvaged from the ash and, in a few cases, the ash is segregated by size. Medium-sized residue may be used for the construction of road beds. Interest has recently [23] been shown in treating the residue as an ore and recovering trace metals but such operations are complicated by the small supplies of the "ore" at any potential treatment site. In many cases, however, the residue is so poorly burned that it is a liability and not an asset. It is not uncommon in coastal cities to find colonies of sea gulls living off the residue pile. At present most of the residue in the U.S. is landfilled.

HEAT RECOVERY AND ENERGY GENERATION

Overall Economics

From the standpoint of overall cost the use of water-wall combustion chambers becomes attractive only if the recovered energy can be sold. The problems in selling the steam generated by an incinerator-boiler are those of finding a market adjacent to the plant and of ensuring a regular supply of steam power. The amount of steam that can be recovered from incineration is about 2.5 lbm steam per lbm of typical refuse for a furnace operated with 60-percent excess air; but the amount recovered will decrease with increasing amounts of excess air and with increase in moisture content of the refuse. Approximately 25 percent of the steam generated can be used internally in the incinerator plant, and the balance is available for sale either directly or after conversion to electrical power. The economics of in-plant generation of electricity for sale are not attractive partly due to the low reliability of incinerators relative to utility boilers and partly due to the small electrical generating capacity of even a large incinerator (40 megawatts for a 2,000-ton/day plant).[4] Certain European units, however, generate electricity for sale and it is claimed [19] that, with the high cost of fuels in Europe, it is economical to do so.

Combined Plants

If electrical generation is to be considered, the economics are most favorable when the refuse can be burned in combination with conventional fuels in a single installation. Several possibilities exist.

1. Two separate water-wall combustion chambers, one for refuse and one for a conventional fuel, discharging their gases to common superheaters, economizers, APC system, and stack, as practiced in Germany.
2. Two separate furnaces with the steam from the refuse unit delivered to superheaters located in an adjoining conventional utility boiler, a scheme that had been accepted by the Pacific Gas and Electric Company in the San Francisco area before the city opted for an alternative disposal system.
3. A single combustion chamber in which shredded refuse (2-inch top-size) is burned largely in suspension together with conventional-fuel-fired burners supplying most of the energy; this last alternative is being evaluated at the Merrimac Station of Union Electric in the city of St. Louis partly as a means of alleviating the low-sulfur-fuel shortage. (Suspension firing of industrial solid wastes, liquids and sludges is also to be used in a 400-ton-per-day incinerator in the Eastman Kodak plant in Rochester, New York). With this approach a small grate is required to burn out larger refuse chunks.

In dual refuse/conventional-fuel installations the fraction of the total heat release provided by the refuse is small enough that any fluctuations in the quality or any disruption in the supply can be readily made up by the conventional fuel, thus overriding one of the major disadvantages of the use of refuse as a fuel.

Sale of Steam

The sale of the heat recovered in incinerators as steam can be limited by the availability of

---

[4] A 40-megawatt plant requires almost the same amount of skilled supervision as a generating plant of 1,000 megawatts. – Editor.

TABLE 7.4

INCINERATORS WITH MAJOR HEAT–RECOVERY OPERATIONS [1]

Location	Type of Installation	Design Refuse Capacity TPD	Refuse Higher Heating Value (Btu/lb)	Type of Heat Recovery System
Atlanta, Georgia	Volund	700	3900[b]	Two-drum bent tube boiler located in each of four furnace flues
Chicago, Illinois (Southwest)	Refractory	1200	–	4-convection boilers
Miami, Florida	Refractory	900	–	4-convection boilers
Hempstead, L.I., N.Y. (Oceanside)	Refractory	600	4500[b]	2-convection boilers
U. S. Naval Station Norfolk, Virginia	Waterwall	360	4700[c] 3840[c]	2-waterwall systems
Braintree[a]	Waterwall	240	5000[b]	2-waterwall systems
Providence, R.I.	Refractory	–	–	4-convection boilers
Oyster Bay, N.Y.	Refractory	–	–	4-convection boilers
Boston, Massachusetts	Refractory	–	–	3-convection boilers
Hempstead, L.I., N.Y. (Merrick)	Refractory	700	4500[b]	2-convection boilers
Chicago, Illinois[a] (Northwest)	Waterwall	1600	5000[b]	4-waterwall systems
Hamilton, Ontario	Waterwall	600	5000[b]	2-waterwall systems
Montreal, Quebec[a]	Waterwall	1200	5000[b]	4-waterwall systems
Rotterdam	Waterwall	1540	2650[c]	4-waterwall systems
Munich (Nord)	Waterwall	1300	2210[c]	2-waterwall systems
Paris (Issy)	Waterwall	1800	3450[c]	4-waterwall systems
Amsterdam	Waterwall	2000	3600[b]	4-waterwall systems
Dusseldorf	Waterwall	1000	2600[b]	4-waterwall systems
Mannheim	Waterwall	1000	2700[b]	2-waterwall systems

Notes: a. Under construction; b. Design value; c. Measured value.

Consumer	Steam Pressure psig	Steam Generation Rate		Steam Generating Efficiency	Steam Credit S/1000=	Design % Excess Air
		= Steam/= Refuse	= Steam/10'' Btu			
Sold to local steam heating system	175	1.62[b]	415[b]	42.0[b]	0.20	–
–	–	2.0[b]	–	–	0.625	–
Sold to a hospital	250	0.330[b]	–	–	0.70	–
2500 kw internal power generation and water desalination	460	2.75[b]	610[b]	61.0[b]	–	150
Used in conjunction with main stream line to supply steam to docked ships	267[c] 280[c]	3.35[c] 2.77[c]	700[c] 722[c]	72.2[c] 71.2[c]	– –	50 –
–	–	3.0[b]	600[b]	–	–	50
–	–	–	–	–	–	–
Internal heating and electric power generation	–	–	–	–	–	–
Partially used for internal heating – steam supply to nearby hospital has been discontinued	–	–	–	–	–	–
–	–	2.0[b]	445[b]	–	–	200
–	250	3.3[b]	660[b]	67.0[b]	–	80
–	–	4.2[b]	–	–	–	50
–	–	4.0[b]	–	–	–	–
Power station	400	1.69[c]	645[c]	72.5[c]	–	–
Power station	2,920	1.58[c]	715[c]	71.5[c]	–	–
Power station	925	2.10[c]	615[c]	72.0[c]	–	–
Power station	611	2.15[b]	595[b]	69.0[b]	–	–
–	1,422	1.6[b]	615[b]	69.0[b]	–	–
–	1,720	1.9[b]	700[b]	69.0[b]	–	–

Courtesy Arthur D. Little, Inc.

a reliable market adjacent to the plant and by the difficulty in ensuring a regular supply of steam. In addition to the potential sale to utilities mentioned above, steam from incinerators may be used for heating large federal, municipal, or private installations and has, in the Oceanside, New York, incinerator been used for supplying the plant's water requirement by desalination [20]. It should be recognized, however, that the energy is a by-product of the incineration process, and its supply cannot be regulated as closely as that of conventional sources and therefore customers must often maintain standby facilities. Auxiliary firing of fossil fuels can smooth the fluctuations in steam-generation rate as has been found to be practiced at the Braintree, Mass. incinerator. Under fluctuating conditions, however, the only credit that can be taken for steam generated by incinerator boilers is the cost of fuel saved by not operating the standby units. The existence of a steam network, such as the network of 90 miles of steam pipes in Paris, into which the steam from incinerators can be fed, eliminates the dependence on any one unit and increases the value of the steam generated by incinerators. With a potential of recovering 2.5 lbm of steam per lbm of refuse and for a product value of \$0.25 to \$0.50 per 1,000 lbm, gross credits up to \$2.50 per ton of refuse can be expected from the sale of steam.

Table 7.4 presents a compilation of units designed for heat recovery, with selected design or performance statistics. It can be seen that major heat recovery is practiced only under special circumstances.

The operation of boilers requires personnel with specialized skills so that consideration of heat recovery by municipalities should be accompanied by a willingness to provide adequate funds to obtain qualified personnel to operate and maintain the more sophisticated equipment.

AIR-POLLUTION CONTROL

An essential requirement of modern incinerators is that they meet the most stringent air-pollution codes. The primary concern with emissions from incinerators has and continues to be the particulate matter emitted, which although relatively small on a national basis has been estimated to account for as much as 20 percent of the total ambient loading in New York City (this is undoubtedly an unusually high value, however). The emission of combustible pollutants, such as carbon monoxide, hydrocarbons, and soot may be controlled in well-designed units by improvement of the combustion process within the furnace itself. The emission rates of other gaseous pollutants such as nitrogen oxides, hydrochloric acid, and sulfur dioxide, are small and have not, so far, presented a problem. Table 7.5 summarizes the expected furnace emissions, before APC, for an "average" incinerator in the U. S.[5] Future changes in refuse composition or operating practise may, however, shift the priority from particulates to some of the volatile compounds, particularly HC1.

Historically, the only codes incinerators had to meet were Ringelmann Smoke Density tests, although many units were engineered to meet the more quantitative 1949 A.S.M.E. code of 0.85 lbm of solids per 1,000 lbm of flue gas corrected to 50 percent excess air. The 1949 A.S.M.E. code corresponds approximately to 9.0 lbm of fly ash per ton of refuse, or to a collection efficiency, on the average, of 55 percent. More stringent codes are expected as a

[5] Naturally, the emission rates for any particular system can vary widely from this average both in respect to the total weight and the distribution among the various pollutants. For example, the particulate emission increases greatly with ash content of refuse. An extreme example is found in Germany, where in winter some units are fired with refuse containing as much as 60 percent ash resulting from use of domestic coal stokers and furnace emissions as high as 130 lbm/ton of refuse are measured.

result of the Clean Air Act of 1967 and the 1970 modifications to the Act. Although codes will vary from state to state it is expected that the upper bound on particulate emission will be in the range of 2 to 4 lbm per ton of refuse.[6] To meet these codes, collection efficiencies of 80 to 90 percent will be required. In many communities, efficiencies up to 99 percent are being sought by concerned citizens.

TABLE 7.5
TYPICAL FURNACE EMISSION FOR AN "AVERAGE" INCINERATOR* [1]

Pollutant	Emission Factor (lbm/ton refuse)
Mineral particulate	15.1
Combustible particulate	4.6
Carbon monoxide	34.8
Nitrogen oxides (as NO_2)	3.0
Hydrocarbons	2.7
Polynuclear hydrocarbons	0.005
Sulfur oxides**	3.94
Hydrogen chloride**	0.99
Lead oxide	0.03

*"Furnace emissions" represents the inlet loading to air-pollution-control devices – in contrast to "Stack emissions" which contribute to air pollution.

**These are maximums for typical refuse. Changes in refuse composition will change these values. Also, some data show these acidic gases to be partially absorbed on alkaline particulate matter.

Gas-quench System

The gases processed by bag filters, mechanical cyclones or electrostatic precipitators have to be cooled to 450F-700F. In water-wall units this cooling is achieved by indirect heat exchange with water. For refractory-wall combustion chambers, the gases may be cooled by dilution with air, or by vaporization of water. Air quenching is impractical in modern units since the flue-gas volume to be treated is approximately doubled in the process. Water quenching, however, results in an approximately 40-percent decrease in volume, since the decrease in temperature on water vaporization more than offsets the increase in mass of the flue gases. The amount of water evaporation required to reduce the gas temperature to the desired range, for a typical refuse burned with 100-percent excess air, is 2 to 2.5 lbm per lbm of refuse. Two designs, wet-bottom and dry-bottom, prevail for the water-quench system. In the wet-bottom

[6]Examples of newer statutes include: (1) the 1966 Executive order 11282 for the control of air pollution originating from Federal installations which imposes a limit of 0.2 grains of particulate per standard cubic foot of dry flue gas corrected to 12-percent carbon dioxide (approximately 4 lbm/ton) for installations burning 200 pounds of refuse or more per hour. (2) State of New Jersey regulations imposing a limit of 0.1 grains of particulate per standard cubic foot of dry flue gas corrected to 12-percent carbon dioxide (2 lbm/ton).

system, water in amounts exceeding twice that required for cooling is pumped through banks of coarse-spray nozzles located in the flue leading to the APC system. The excess water collects in the bottom of the flue, and must be treated before disposal or recycle. The treatment of the water stream, which is highly acidic (pH<2.0) due to the absorption of HCl, SO_x and organic acids, and which is contaminated with particulates, is a major disadvantage of the wet-bottom system. It is, however, cheaper to install and easier to operate than the dry-bottom system in which the water is injected, in the amounts required, as a fine enough spray that complete droplet evaporation is achieved. To provide the residence time required for evaporation, cooling towers 30-feet to 90-feet high are provided. The dry system requires high water pressures, from 100 to 600 psig, and filtered water to prevent clogging of the fine spray nozzles.

Since the size and cost of APC systems scale with volumetric flowrate of gases to be treated, it is of interest to note the relative volumes of gases emerging from different quench systems. For an average ton of refuse burned with 50-percent excess air in a water-wall unit, the volume of gases delivered to the APC system at 500F is about 270,000 cubic feet, for a refractory-wall combustion chamber, 100-percent excess air and a water quench to 500F the volume is about 560,000 cubic feet; and for a refractory-wall unit, 100-percent excess air and an air quench the volume of gases delivered at 500F to the APC system is about 1,800,000 cubic feet. The savings to be had by operating with low excess air and by eliminating or minimizing the addition of quench fluid are significant.

Particulate Emission and Control

In controlling the emission from an incinerator, the possibility of reducing the emission from the combustion chamber, and thus decreasing the requirements of the APC system, should be thoroughly investigated. Reduction of undergrate air velocity and/or bed agitation reduces particulate entrainment from the bed; the proper design of overfire jets reduces the emission of combustibles.

With wide variations in particulate loading from combustion chambers it should be evident that an APC system should be tailored to the emission of a particular installation rather than to the average values cited in Table 7.5. Another important factor which must be taken into consideration is the particle-size distribution of the fly ash leaving the furnace enclosure.

Particle-size distributions of fly ash obtained on three U.S. incinerators [21] are compared with average values for European incinerators in Figure 7.5. On the basis of the limited data in the figure it is apparent that the attainment of high-efficiency collection, say 95 percent, will be more difficult in U.S. incinerators since the emission from furnaces in the U.S. apparently contains much higher percentages of fine particles. The strong dependence of collection efficiency with particle size for most APC systems is illustrated in Table 7.6, taken from [22]. It is apparent that bag filters, electrostatic precipitators, and high-pressure-drop scrubbers are the only systems that yield high efficiencies for particles smaller than five microns.

A rough guide of performance and relative cost of installation and operation of different APC systems is presented in Table 7.7, based on figures prepared by Niessen et al. [1]. The wet settling chamber and mechanical cyclones are excluded from consideration for new installations on the basis of performance. The three systems remaining that appear to meet projected air-pollution legislation are the electrostatic precipitators, bag filters, and scrubbers. The last has the advantage of absorbing highly soluble gases such as HCl and the disadvantage of producing a steam plume and of requiring a water-treatment system. The trend in U.S. practice

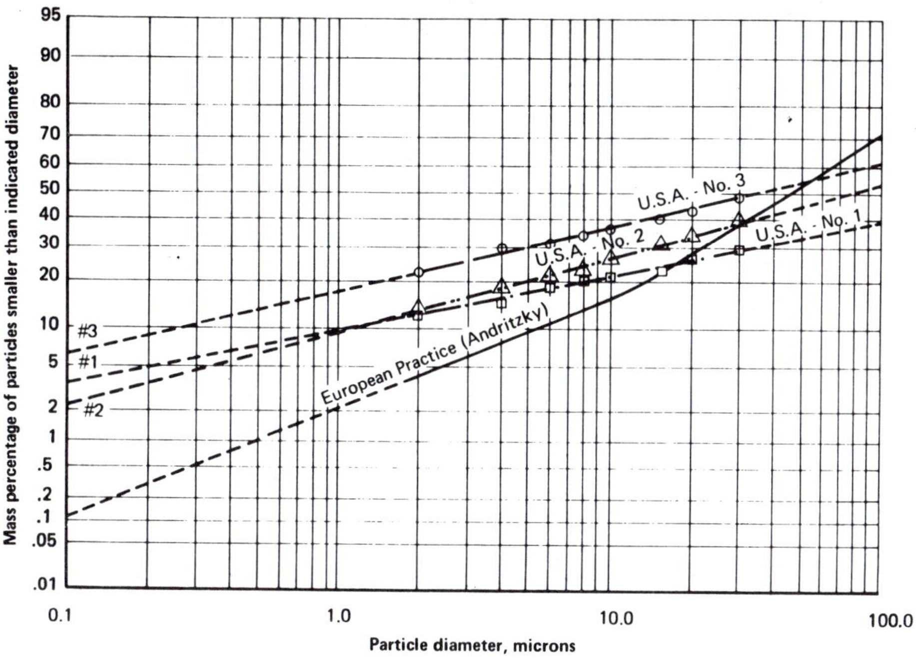

Figure 7.5 Particle-size distributions of furnace effluents.

courtesy Arthur D. Little, Inc. [1]

since 1940 has been towards more efficient APC systems (see Figure 7.6), primarily wet-collection methods or cyclones. Electrostatic precipitators are commonly used on large municipal units in Europe, have been tested on pilot-plant units in New York City, and have been introduced on a full-scale basis in the U.S. and Canada in several plants.

DESIGN

The requirements and specifications for incinerators for each community are such that no two incinerators are built alike. Consulting engineers usually select the component elements and, together with the equipment vendors and prime contractors, establish the interfacing between elements, usually under severe restrictions of financing and local regulations. *The responsibility for the total system design is so segmented that no single group has responsibility or can be held accountable for over-all system performance.* This arrangement often has serious consequences.

Criteria of Performance

The factors which should be considered in the selection of an incinerator are reliability, residue quality, environmental pollution, safety, and cost. It is, however, difficult to establish quantitative criteria for these.

The reliability of a unit must be geared to the total disposal problem in an area. Where there

TABLE 7.6
EFFICIENCY OF DUST-ARRESTING EQUIPMENT

Equipment	Percentage efficiency at 50μ	5μ	1μ
Inertial collector	95	16	3
Medium-efficiency cyclone	94	27	8
Cellular cyclone	98	42	13
High-efficiency cyclone	96	73	27
Tubular cyclones	100	89	40
Jet-impingement scrubber	98	83	40
Irrigated cyclone	100	87	42
Self-induced-spray deduster	100	94	48
Spray tower	99	94	55
Fluidized-bed scrubber	> 99	98	58
Irrigated-target scrubber	100	97	80
Electrostatic precipitator	> 99	99	86
Desintegrator	100	98	91
Irrigated electrostatic precipitator	> 99	98	92
Annular-throat scrubber—low energy	100	>99	96
Venturi-scrubber—medium energy	100	>99	97
Annular-throat scrubber—medium energy	100	>99	97
Venturi-scrubber—high energy	100	>99	98
Shaker-type fabric filter	> 99	>99	99
Low-velocity bag filter	100	>99	99
Reverse-jet fabric filter	100	>99	99

is no alternative to incineration for disposal, allowance can be made for unscheduled shutdowns by using two or more furnaces. Similarly, furnace down time can be minimized by maintaining spare parts of key components and by providing by-passes in the system.

The residue quality should satisfy the constraints on volume reduction, putrescibles, combustibles, water pollution, and odors imposed by the method of disposal of the residue. Completion of combustion should exceed 97 percent. Unfortunately, however, many units are operated at burning rates exceeding design conditions with a consequence that much of the refuse may be only singed in the incinerator. Requirements of both designer and plant superintendent to aim for a given percent burnout is desirable – but it is difficult to monitor burnout.

In addition to the air-pollution problem discussed earlier, odors, dust, water, noise and aesthetic pollution need be considered.

Safety requirements for incinerator design should be similar to those for industrial plants dealing with high-temperature operation. Periodic explosions due to receipt of batches of rapid-burning wastes should be anticipated.

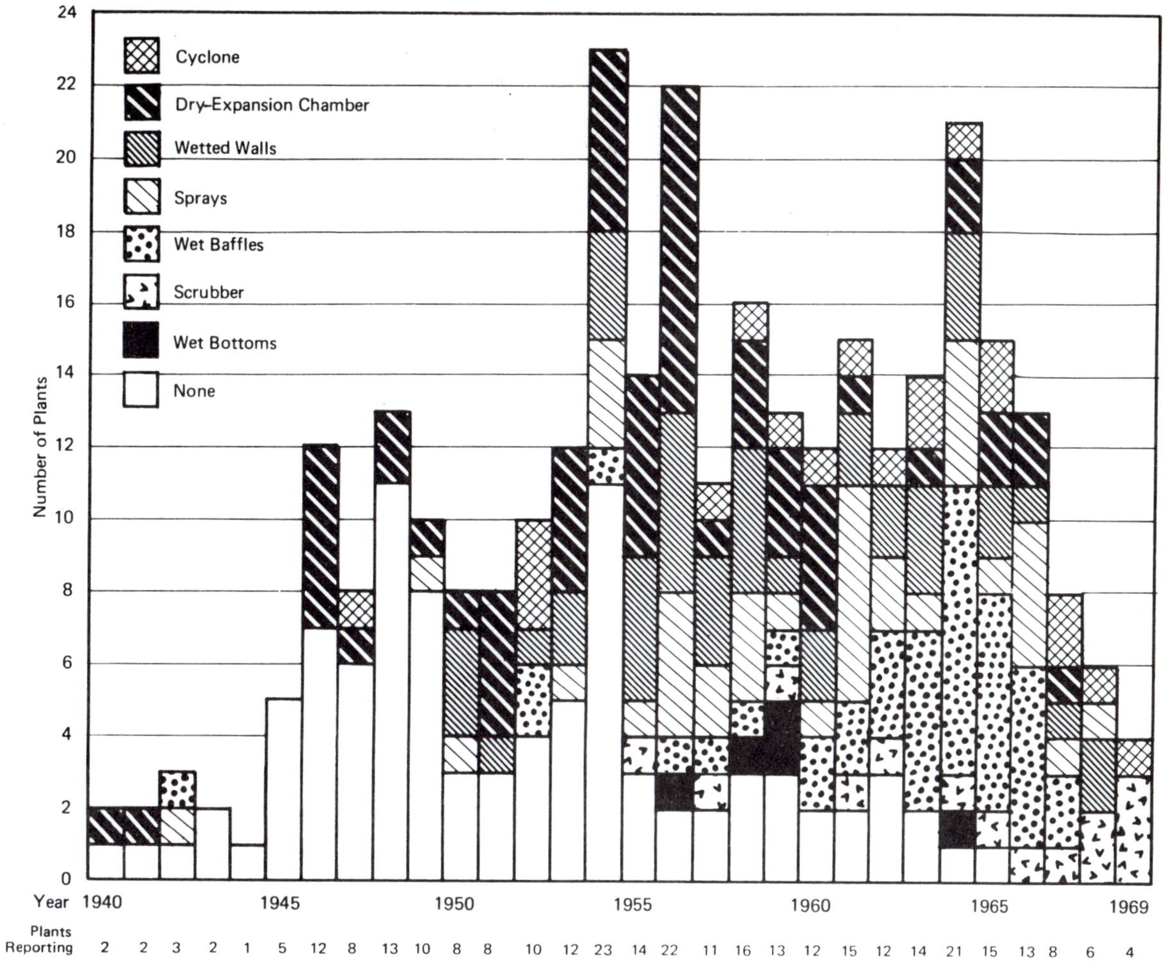

Figure 7.6 Primary air-pollution-control equipment.

courtesy Arthur D. Little, Inc. [1]

Incinerators are often judged on first cost, and the high cost of operation resulting from an inferior design is overlooked. In the evaluation of bids it is the total operating and amortization costs that should be of interest. For this reason the Bureau of Solid Waste Management has been trying to encourage the adoption of a standardized cost-accounting system [6].

Sample Designs

As mentioned above, practically every incinerator is custom-designed. Attempts to realize the economies to be gained from repeat sales of a fixed design have only recently been made. Combustion Engineering has designed one such unit with a capacity of 75 TPD for small communities. Figure 7.7 shows a schematic of the Combustion Engineering unit together with those for intermediate-scale and large-scale plants partly to show how the design varies with scale (batch vs. continuous, refractory vs. water-wall, scrubber vs. electrostatic precipitator), and partly to illustrate how the component parts discussed in the previous sections are fitted together.

TABLE 7.7
TOTAL INSTALLATION COSTS, ANNUAL OPERATING COSTS AND EFFICIENCIES FOR SEVERAL APC SYSTEMS

APC system	Total installed[1] cost, $/CFM	Annual operating[2] cost, $/CFM	Collection efficiency, percent
Wet settling chamber	0.28	0.028	40
Cyclone ($\Delta P = 3.5''H_2O$)	0.60	0.12	68
Cyclone ($\Delta P = 5.0''H_2O$)	0.72	0.17	83
Scrubber ($\Delta P = 5.0''H_2O$)	1.10	0.41	91
Scrubber ($\Delta P = 15.0''H_2O$)	1.70	0.65	93
Electrostatic precipitator (95% Eff.)	1.65	0.24	95
Electrostatic precipitator (99% Eff.)	1.30	0.19	99
Fabric filter	1.75	0.32	99.5

[1]Estimated for a 400-TPD unit (≃ 200,000 cfm) using the 1969 Cost Index for the Boston area. The costs will vary with size, efficiency and location. In particular the costs of electrostatic precipitators increase more rapidly than the other systems as the unit size is decreased.

[2]Estimated for a 400-TPD unit in the Boston area operating 3 shifts/day, 6,000 hrs/year.

Source: Reference 1

Operation and Control

Despite the complexity of operating an incinerator efficiently, municipalities have not in general employed skilled labor. As a consequence of poor working conditions many incinerator owners have difficulty in keeping even their unskilled labor and absenteeism is high. But even in installations in which the operators are highly motivated, they may be limited in their performance by a lack of training and by the restricted funds for maintenance. Although most recent incinerator units are equipped with control panels and gauges these are too often rarely referred to or adequately maintained. The problems of overcoming the stigma of working with "garbage" are enormous and are a serious handicap to advancement in incinerator sophistication in the U.S. One example of the problem is the difficulty that consulting engineers are encountering in persuading municipalities that a water-wall incinerator generating steam requires supervision by a boiler engineer. Until incinerators are recognized as being process plants and staffed accordingly, the operation and control of the units will continue to be guided more by instinct and experience than by an appreciation of the engineering principles involved.

American Practice

The total active incinerator capacity in the U.S. is approximately 100,000 TPD which, for an estimated use factor of 50 percent, accounts for the disposal of a little under ten percent of the municipal and light commercial solid waste generated in the United States. It is expected,

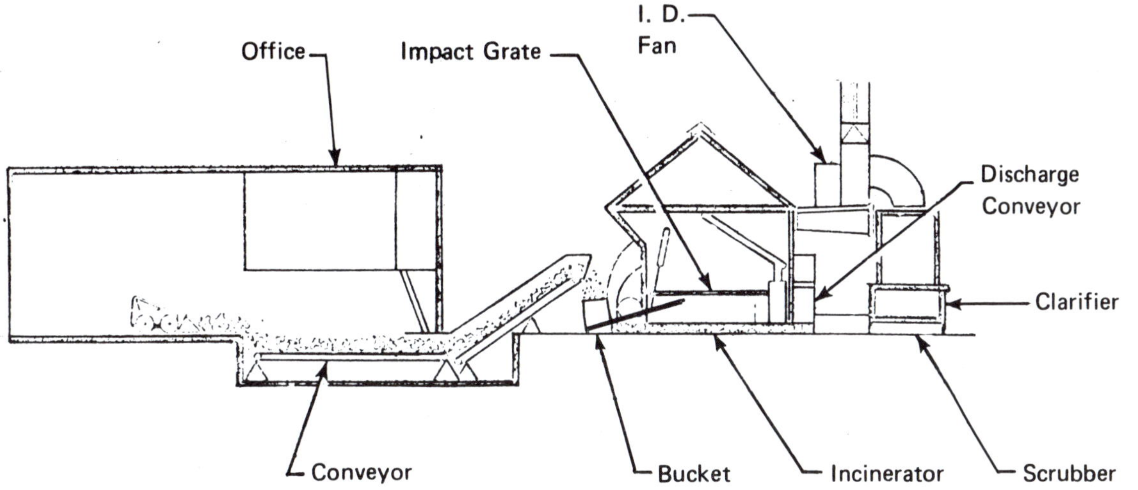

Figure 7.7a Combustion-engineering package unit.

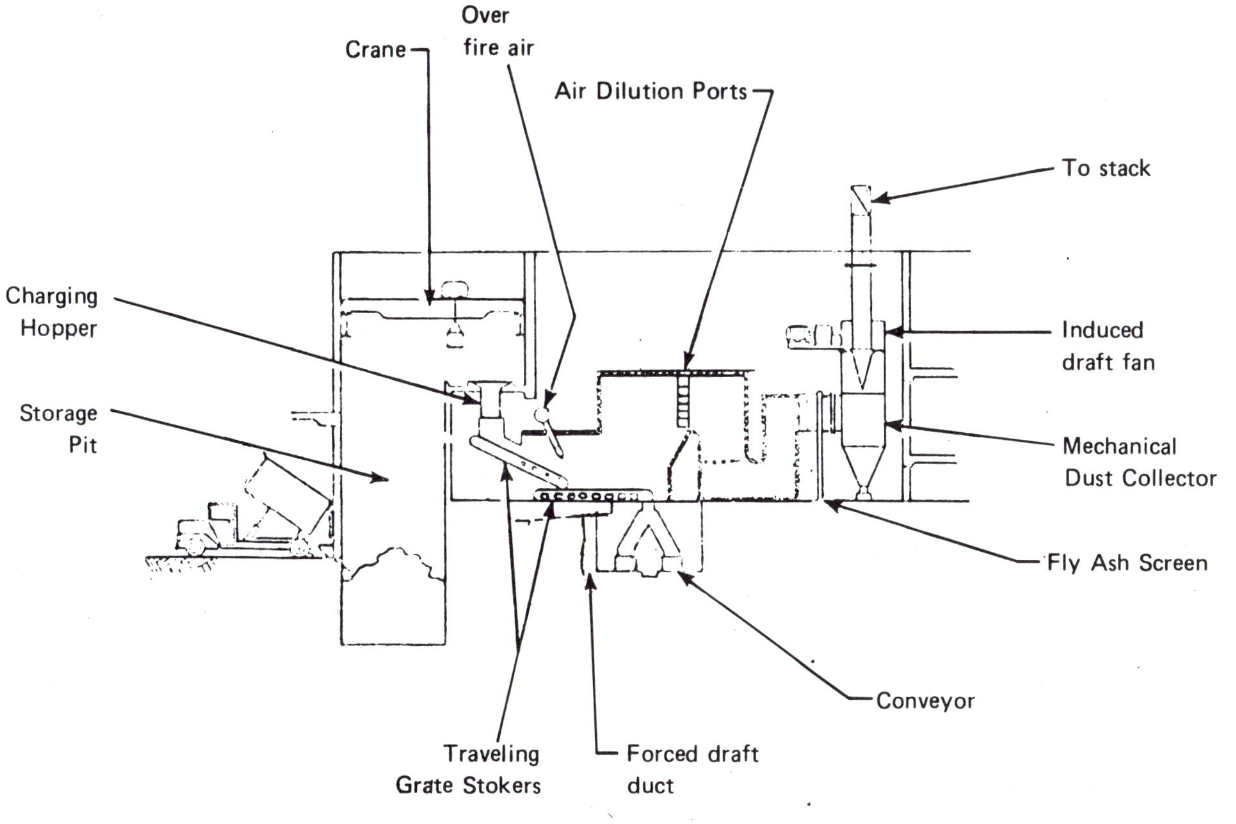

Figure 7.7b Refractory-wall unit [25].

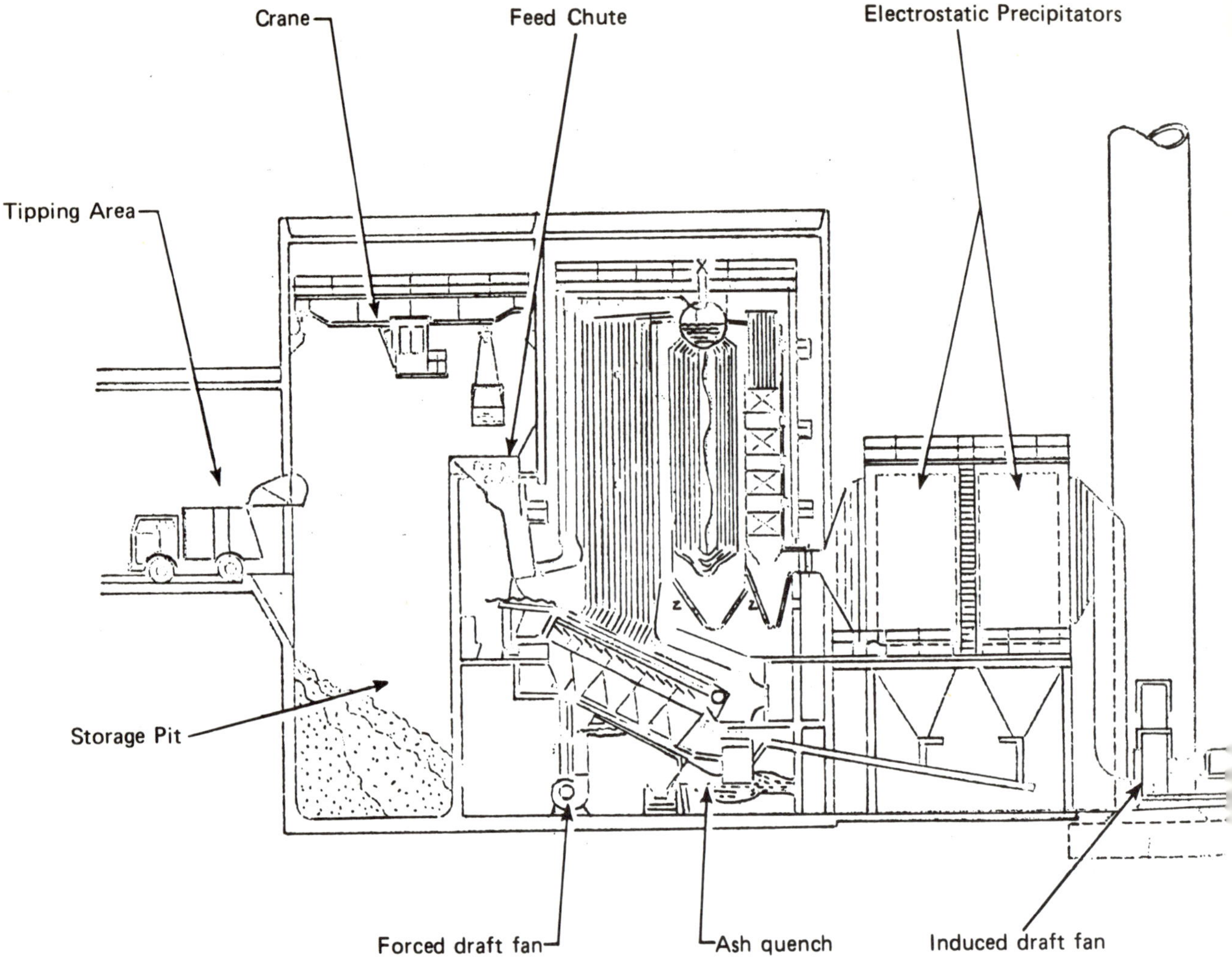

Figure 7.7c Water-wall unit [24].

however, that incineration accounts for a much larger fraction of the waste in highly populated urban areas. For example, in New York City, 2,254,000 tons or 35 percent of the collected household, commercial, and government waste were incinerated by *municipal* units alone in 1968. In other regions tradition or other considerations may influence the selection of disposal method. For example, in Wisconsin an unusually large number of small municipal incinerators are found apparently as a consequence of the traditions brought over from Germany; while in Los Angeles County incinerators are banned as part of the country's pollution-abatement program.

The total annual additions to United States incinerator capacity shown in Figure 7.8 illustrate the relatively small market and the great variations in sales. Of particular interest is the 1962 to 1969 drop which is probably explained by the unwillingness by communities to commit large investments, borrowed at high interest rates, to incinerators at a time when air-pollution codes were being revised and many new, supposedly inexpensive, high-

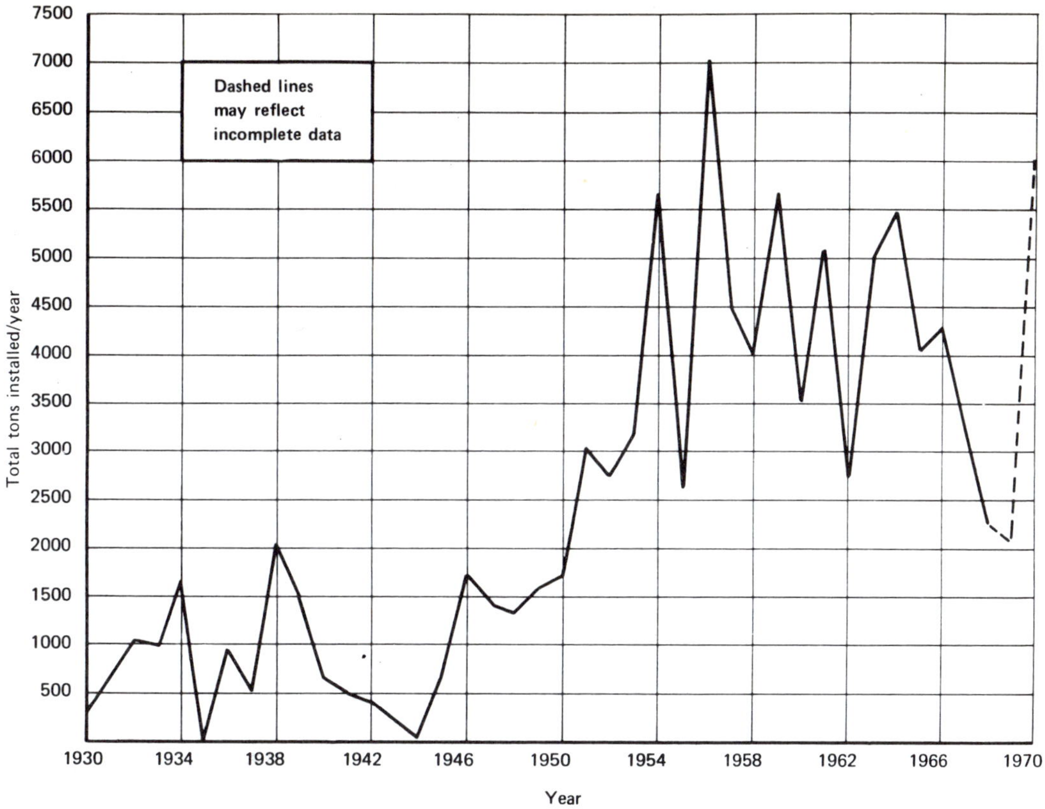

Figure 7.8 Total annual additions to United States incinerator capacity.

courtesy Arthur D. Little, Inc.

performance waste-disposal concepts were being promoted. Due, in part, to this "pause," the waste-disposal problem in many communities has reached critical proportions while officials try to respond to public pressures for an upgrading of disposal facilities without incurring any additional costs.

THE ECONOMICS OF CONTEMPORARY INCINERATION SYSTEMS

Economic decisions regarding incinerator design efforts, capital investment, and operating expenses are often the key to the success or failure of incinerator plants. Indeed, the succinct homilies which follow could be the epitaph of many, if not most, U.S. incinerators:

- A stitch in time saves nine (inadequate design budgets)
- You get what you pay for (inferior or inadequate equipment)
- Penny-wise and pound-foolish (little maintenance, unskilled labor)

In comparison with most other alternatives in solid-waste disposal, incineration is a costly process which can be made still more costly by the inefficiency, unreliability, air and water pollution, ugliness and so forth which are consequences of such poor economic judgments. Municipalities selecting incineration as an optimal means of meeting a pressing solid-waste-management crisis must accept the responsibility for doing the job correctly. If it is argued that incineration is too expensive, perhaps the original thinking and decision-making processes which suggest incineration as the optimal course should be reexamined. It should be interjected here that a first-class plant does not mean "gold-plated" but merely a system offering performance which reflects modern technology and a full measure of public responsibility.

Review of Economic Factors

The growing demand for new solid-waste-disposal facilities throughout the country stresses the need for a greater understanding of the available technology for refuse incineration and, in parallel, an appreciation of the economics associated with the various concepts. The paragraphs below present a review of the economic considerations.

When selecting an incineration system, there are different and often competitive criteria by which performance may be judged. The principal criteria applicable to incineration – reliability, residue quality, environmental pollution, safety and cost – must be carefully considered in determining the performance levels which a municipality desires from its incineration facility. For example, if the air-pollution-control regulations are stringent and plans have been formulated to use the incinerator residue for road-bed applications, the municipality should strongly consider designs offering highly efficient combustion systems and effective air-pollution-control devices. Such environmental-control and residue-quality requirements must be considered in view of the ever-present criterion of cost to determine the type of system best suited for the municipality.

The paragraphs below provide rough economic information for contemporary batch-fired and continuous-fired incineration systems. In presenting the overall plant costs, all plant types were evaluated for the same air-pollution-control (APC) system, a high-efficiency electrostatic precipitator. It should be noted that this evaluation is not at the same level of stack emission for different incinerator types because of differences in the emissions entering the APC device.

There are severe limitations to the direct application of the economic data presented subsequently for any specific location. Although [1] suggests ways to modify these and other economic data to conform with localized cost factors (using the Engineering News Record (ENR) indices), there are a number of other important considerations not readily incorporated into such estimates. Investment costs, for example, assume "average" foundation costs. If the load-bearing capability of the local soil is unusually low, pilings or other extensive foundation work will add significantly to the cost. Similarly, if the plant is built on bedrock, the blasting which may be required for the refuse-storage pit will also add significantly to the cost. The primary utility of these estimates, therefore, lies in comparing concepts and in visualizing the relative importance of various cost factors in the total operating-cost evaluations.

Capital Investment

To develop the generalized investment costs shown below, data from a number of sources were collected and analyzed. The principal data sources were the consulting engineering firms and equipment vendors. Also data collected by the Bureau of Solid Waste Management and published information were used. From this data bank, broad ranges of costs were outlined, reflecting in part the diversity of equipment options and also differences in local construction costs. The capital-investment presentations for batch-feed and continuous-feed construction systems show the range of costs which were identified. These ranges are somewhat smaller than the raw data indicated, as the effects of year and location of construction have been partially removed by use of the ENR indices to bring all costs to a reference point of Boston, Massachusetts in mid-1969. The residual variability in capital-cost estimates, however, is at least ±15-20 percent.

As described above, an incineration concept can utilize a wide variety of sub-systems, each of which may significantly affect plant performance and economics. Feeding systems, for example, may involve a conventional (and expensive) refuse storage pit and crane, or alternatively, a refuse dumping and storage floor serviced with a front-end loader and slat-conveyor system. Handling systems for either wet or dry residue and siftings can be utilized. Air-pollution-control systems can vary from simple spray and water-impingement-baffle systems to high-energy scrubbers or electrostatic precipitators. The system can be refractory-lined or equipped with waterwalls; flue-gas cooling can be by air dilution, water evaporation, or by heat removal with a convective boiler.

In order, then, to compare the economic aspects of the various incineration concepts now used, a common design basis was selected. The data which follow assume that no waste-heat or metal-recovery credits are taken, that refuse is fed directly from a pit to the furnace by means of a bridge crane and bucket, and that the incineration facility is on land which is suitable for construction without excessive site preparation or access-road construction. The instrumentation and controls utilized are those of typical modern existing incineration facilities.

Batch-feed Incineration Systems

Examination of capital-investment data for existing batch-incineration systems shows a wide variation in the costs of such facilities. The data shown in Figure 7.9 were accumulated for the furnace and building portions of existing batch systems (both of the rectangular and cylindrical type) adjusted to reflect labor and materials costs in Boston, Massachusetts for mid-1969. The appropriate incremental costs for electrostatic precipitators were then included to make up the total investment. It should be noted that in many cases, precipitators would be a technically poor choice unless care was taken in the refuse-feed system to avoid wide swings in flue-gas flow rate (see above). Figure 7.9 shows the resulting plant costs in dollars per ton of rated capacity per 24-hour day (TPD) as a function of the capacity of furnace units comprising a plant with two or more furnaces. The range of typical furnace, building and electrostatic-precipitator costs is shown. The furnace costs include the costs for: the crane, the primary and secondary combustion chambers, forced-draft fans and ducting, residue-removal system, gas-cooling system, breeching and stack, piping, instrumentation and controls. The building costs include excavation, land improvements, foundations, the refuse pits, buildings and structural supports, and a moderate investment in access roads. The electrostatic-precipitator costs include the induced-draft fan plus all auxiliaries, motors and ducting directly associated with the precipitator system.

Figure 7.9 illustrates well the wide range of batch-fed incinerator costs which presently characterize the incineration industry. For example, an incinerator comprised of two, 150-TPD units might generally be expected to have an initial capital cost of about $6,000 per daily ton. However, because of variations in the parameters discussed above, plants of similar capacity could cost from $4,400 to $6,700 per daily ton.

Continuous-feed, with Refractory or Waterwall Construction Incineration Systems

The continuous-feed incinerator with waterwall construction and grate burning is similar in many respects to existing plants with refractory-wall construction. Refuse-feeding, grate, siftings-removal, and ash-handling sub-systems are or can be identical. The primary differences lie in the operating characteristics of the units. For example, since the boundaries of the waterwall system can absorb high heating loads without damage, it becomes practical to

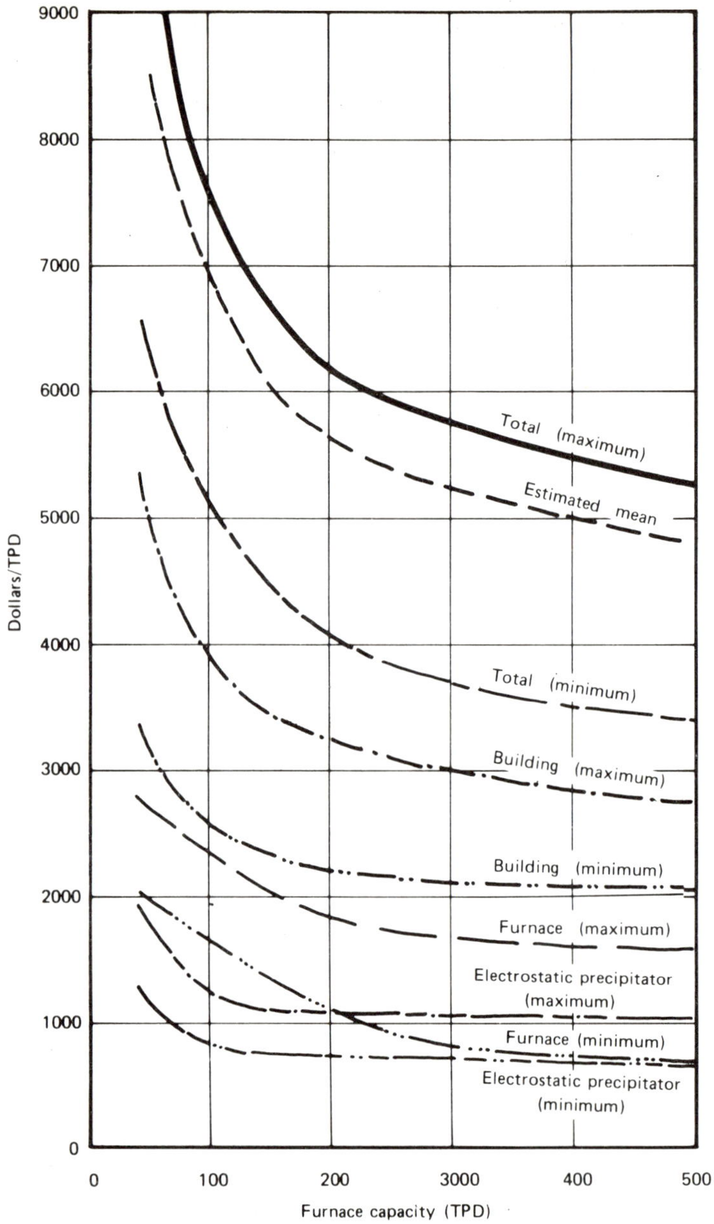

Figure 7.9 Average investment costs: batch-feed systems.

courtesy Arthur D. Little, Inc. [1]

operate at low excess-air levels (50-100 percent) in contrast to most refractory systems, which operate with above 150-percent excess air. Also, most waterwalls are fabricated without seams, reducing the leakage of air into the chamber to that associated with peep holes, stoking doors, and the like. This reduction in overall flue-gas volume decreases the investment cost at the back end of the system by reducing the size and power requirements for the air-pollution-control system, fan and stack.

Figure 7.10 shows the estimated investment costs for refractory construction incinerator plants. Figure 7.11 shows a similar relationship for grate-fired waterwall units with convection

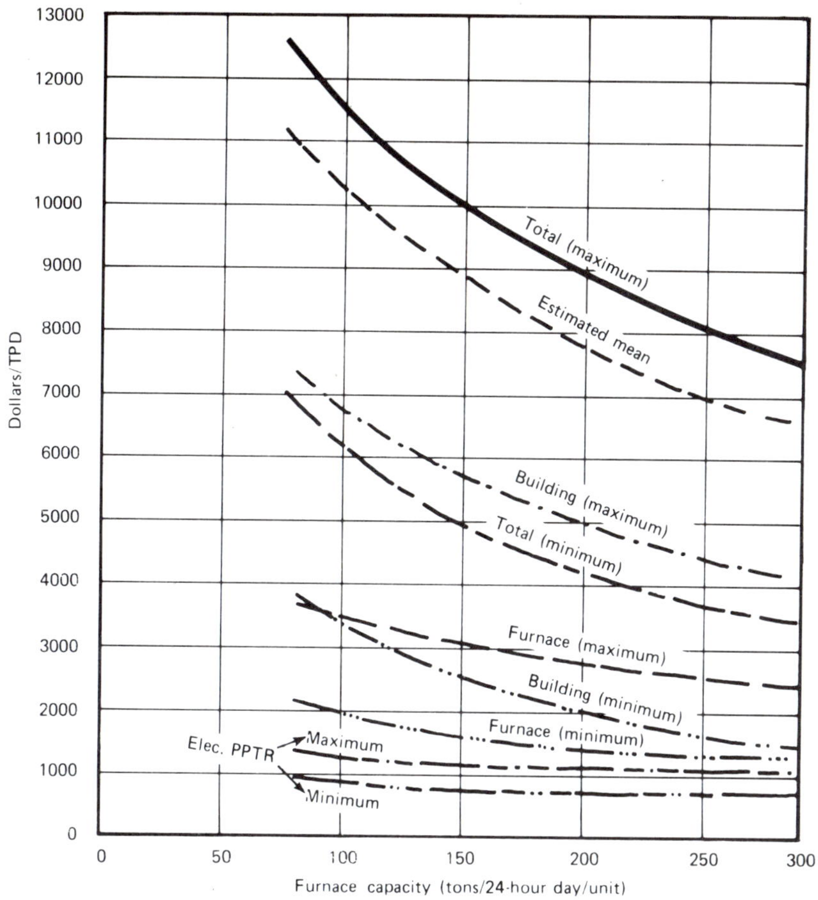

Figure 7.10 Average investment costs: continuous-feed systems.

courtesy Arthur D. Little, Inc. [1]

boilers and electrostatic precipitators. Figure 7.10 was compiled from bid prices, published estimates, and engineering judgments. The values include a ten-percent over-ride for site preparation, costs for a steel stack and a building with housed pit, but no housing over the tipping floor. Since steam demand may not exist during periods of time when the incinerator is operating, the cost for an air condenser has also been included. Figure 7.11 shows estimates of the investment on a comparable basis as a function of TPD per furnace plant. These costs would not change appreciably for multiple-furnace plants.

The two extreme lines on these curves indicate the probable cost ranges which, of course, would be predicated on the details of design and specifications. The ranges as indicated should not in any way be construed as a variation in cost due to geographical location, as all capital costs were based on mid-1969 dollars for Boston, Massachusetts.

Waterwall-construction incineration systems can also be constructed and operated to burn a substantial fraction of the refuse in suspension. This arrangement should reduce the overall furnace volume or could allow larger capacities (say, to 1200 TPD) in a single furnace and it is estimated that excluding capital costs associated with the refuse size-reduction operation, the investment costs for the incineration system will be approximately 15 percent less than those

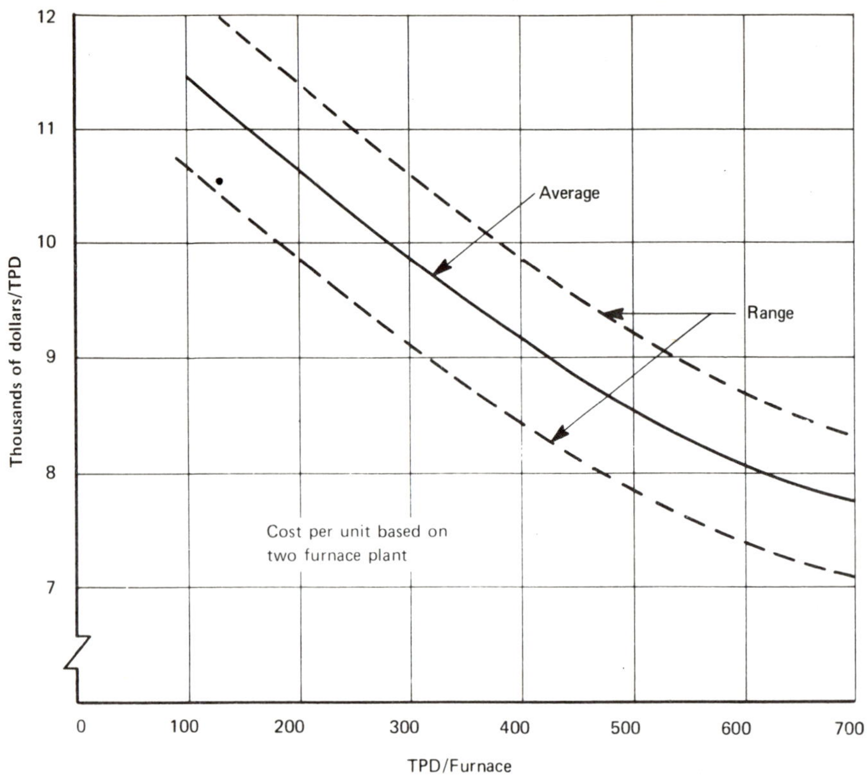

Figure 7.11 Estimated 1969 cost of incinerator plant with boilers and precipitators.

courtesy Arthur D. Little, Inc. [1]

for grate-burning waterwall systems using hopper- and chute-feed methods. Typical costs for installation of refuse-milling equipment, however, would add from \$500 to \$1,000 per TPD to the investment cost of the facility and, as shown below, milling adds significantly to the unit operating cost of a plant.

Package Incineration Systems

The concept of "package units" (prefabricated at a manufacturing facility and shipped substantially ready to operate) has been successfully applied to a number of combustion systems, especially boiler units. Since little field work is required (except for the pouring of an appropriate concrete pad and provision of utility services), the cost and the time from order placement to facility operation are minimized. In addition, since nearly identical systems are provided, engineering costs per unit, and technical risk, are also reduced. A package incinerator is now being offered by Combustion Engineering, Inc, (Figure 7.7).

The cost of the Combustion Engineering package incineration system (installed) has been estimated to be approximately \$250,000 for a single, 72-TPD unit. This includes all equipment, the associated refuse-storage building, site preparation, limited roadways and foundations. The cost for such a system employing an electrostatic precipitator preceeded by a coarse-spray flue-gas-cooling system (in place of the high-performance scrubber unit designed for the system) would cost approximately $360,000, or an additional $1,500 per TPD. (Costs are for Boston in mid-1969.)

Operating Costs

The generalized operating costs shown below were developed using data from a number of sources, collected and analyzed to determine average values for the cost per unit of refuse burned. The principal data sources were consulting-engineering firms, equipment vendors, and Bureau of Solid Waste Management reports.

Reported operating costs can vary widely. The false economy of deferred maintenance, or the penalties of under-designed equipment, for examples, cause wide swings in published values for maintenance expense for similar plants. In some plants, efficient operation is obtained with only three or four operators whereas, in others, unreasonably large crews may be found. In order, then, to compare the economic aspects of various incineration concepts, a common operating basis must be selected. The data which follow assume that no waste-heat or recovery credits are taken; and, as indicated in the tables, standard unit costs for electricity, water, labor and overhead were assumed. A reasonable crew size was taken, one comparable with many of the better-run plants around the country.

Batch-feed Incineration Systems

Figure 7.12 shows the total operating costs (including capital charges) for cylindrical- or rectangular-type batch-feed incineration plants. This figure clearly shows the advantages in multi-shift operation. For example, at a 300-TPD disposal rate, a plant run three shifts experiences a total operating-cost reduction of almost $2 per ton in comparison with one-shift operation and simultaneously acts to reduce furnace maintenance by holding more-constant temperatures, avoiding refractory spalling, condensation and so forth. Table 7.8 shows operating economics calculated for a specific 400-TPD plant to illustrate the breakdown of costs between the various cost centers.

Continuous-feed Refractory-construction Incineration Systems

Figure 7.13 shows costs for continuous-feed systems. It can be seen that in larger plants, the operating cost for continuous-feed systems is only slightly higher than that for batch-feed systems. For a minor additional cost, however, substantially improved residue quality and diminished combustible air-pollutant emissions are obtained. It can be easily seen, however, that the unit cost for continuous systems increases rapidly as plant size diminishes.

Continuous Feed with Waterwall Boiler Construction (grate-burning) Incineration Systems

Figure 7.14 shows the operating cost for waterwall incineration systems. Notice that only three-shift-per-day operation is shown, as continuous operation is highly desirable with boiler systems. It can be seen, in comparison with the continuous-operation refractory systems, that waterwall operation, due to the higher capital investment, is more costly. This illustrates the necessity for steam credits in order to justify waterwall operation. In comparison with refractory systems, for example, a credit of 26¢ per thousand pounds of steam (¢/m lbm) is required for a break-even operation at 1,000TPD capacity. This required credit increases to 46¢/m lbm of steam at 400 TPD. Steam credits in excess of these balancing revenues may be applied against the cost of incineration at an average rate of about 4¢/ton for each cent/m lbm.

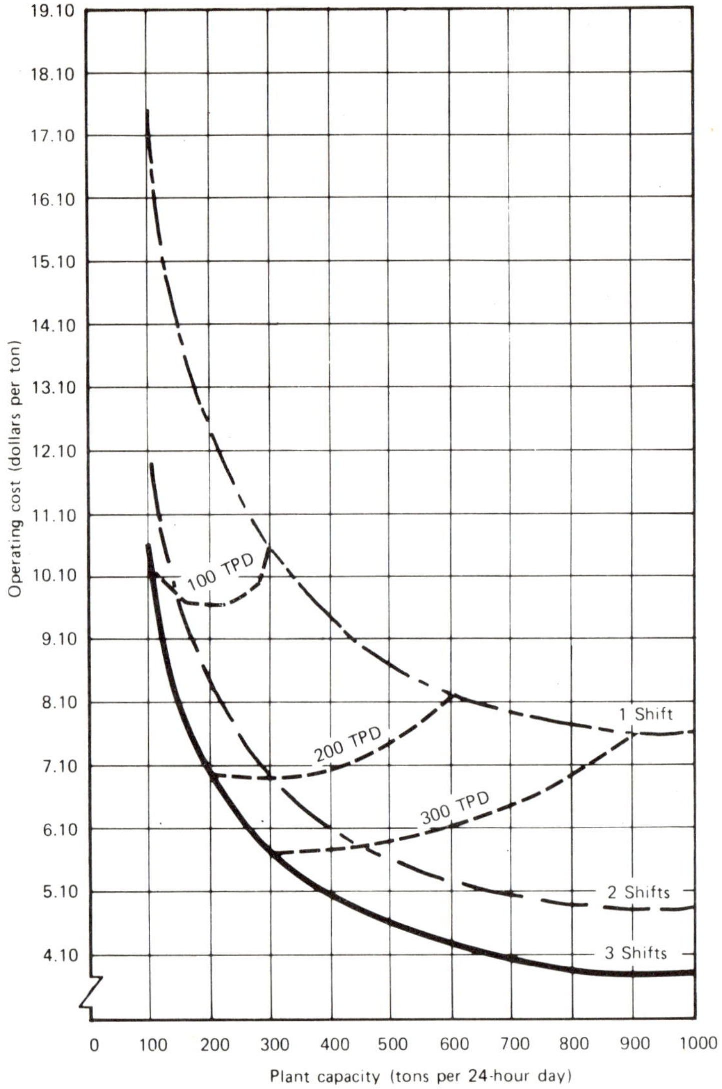

Figure 7.12 Total operating cost versus plant capacity for batch-feed units.

courtesy Arthur D. Little, Inc. [1]

Suspension-burning systems incur an additional cost due to the need for refuse shredding. These costs are shown in Figure 7.15 which, for the two-inch maximum particle size which could be expected for incineration applications, would add approximately two dollars per ton to the incineration cost in a thousand-TPD plant. Slightly lower capital costs associated with suspension-burning systems would partially compensate for the milling cost. The cost increment for suspension-burning systems over refractory furnaces would approximate \$2.80 per ton burned. This suggests the need for a minimum of 90¢/m lbm of steam for a break-even comparison with conventional refractory-wall systems.

Package Incineration Systems

Because of the low capital investments associated with package incineration systems, significant economies in operation can be realized. Table 7.9 shows the estimated operating cost for

TABLE 7.8
ESTIMATED COST OF INCINERATION
USING BATCH INCINERATION SYSTEMS (1969 DOLLARS)

Reference location:	Boston, Massachusetts, mid-1969
Plant capacity:	400 TPD (two, 200-TPD furnaces)
Load factor:	250 days per year; 3 shifts/day
Annual consumption:	100,000 tons refuse/year
Capital investment:	$2,300,000

	Units	Dollars/unit	Units/ton refuse consumed	Dollars/ton refuse consumed
Variable costs				
Electricity	KWH	$0.015	22.3	0.33
Water	M Gal	$0.30	0.90	0.27
				$0.60
Semi-variable costs				
Labor				
1 crane operator per shift @ $5.00/hour				0.31
2 furnace operators (stoking & dumping) per shift @ $4.00/hour				0.38
1 charging-floor man per shift @ $3.00/hour				0.19
2 residue-removal men per shift @ $3.00/hour				0.38
1 supervisor @ $14,000/year				0.14
1 weigh-scale operator @ $5,500/year				0.06
				$1.46
Overhead @ 40% of labor & supervision				0.58
Maintenance and repair @ 2.5% of capital investment annually				0.58
				$2.62
Fixed costs				
Amortization @ 4.7% of capital investment annually				1.08
Interest on loan @ 3.0% of capital investment annually				0.69
				$1.77
Total costs				$4.98

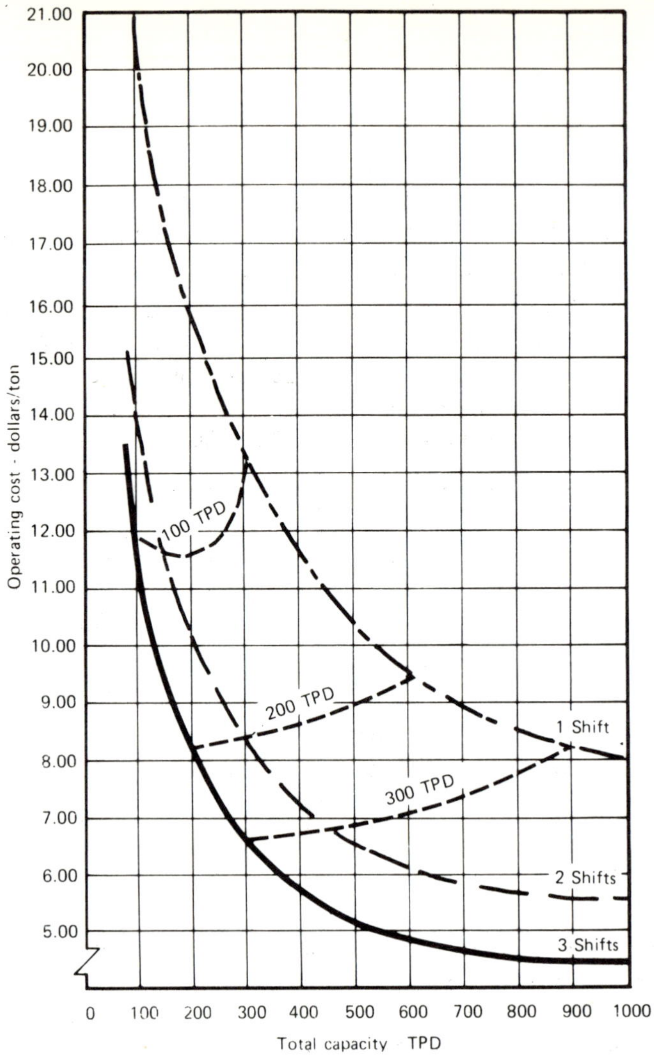

Figure 7.13 Total operating cost versus plant capacity for continuous-feed, rectangular-construction units.

courtesy Arthur D. Little Inc. [1]

TABLE 7.9
THE EFFECTS OF NUMBER OF SHIFTS AND NUMBER OF UNITS ON TOTAL OPERATING COSTS FOR PACKAGE INCINERATION SYSTEMS

Tons consumed per day	No. of units	No. of shifts	Total costs per ton
144	2	3	5.37
96	2	2	6.43
72	1	3	6.75
48	1	2	7.81
48	2	1	9.59
24	1	1	10.97

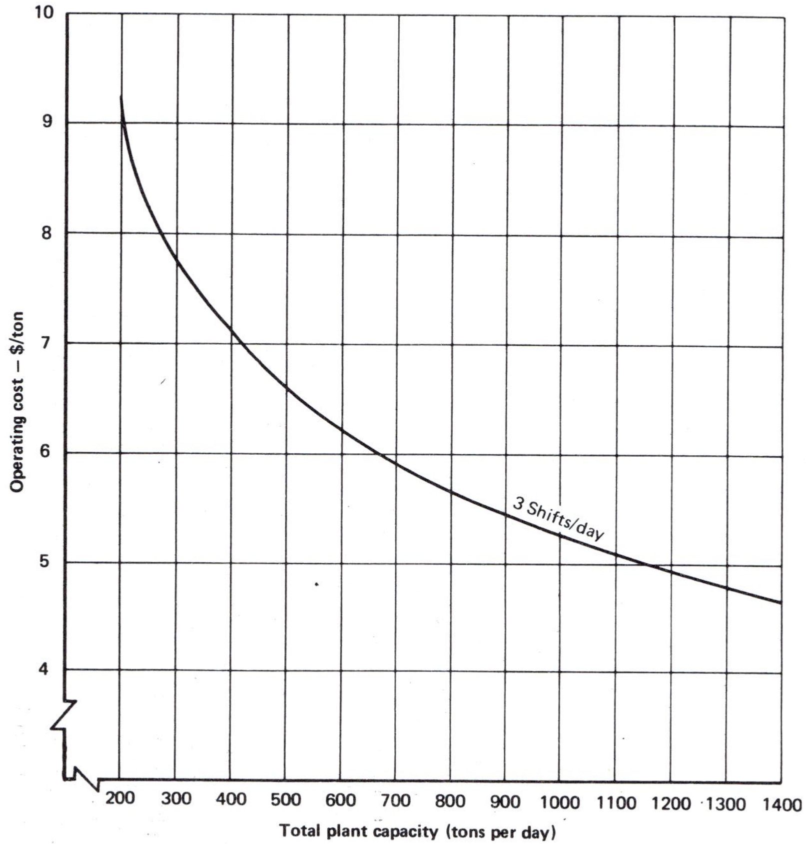

Figure 7.14 Total operating costs—continuous-feed, water-wall-construction incinerator systems (grate-burning, no steam credits).

courtesy Arthur D. Little, Inc. [1]

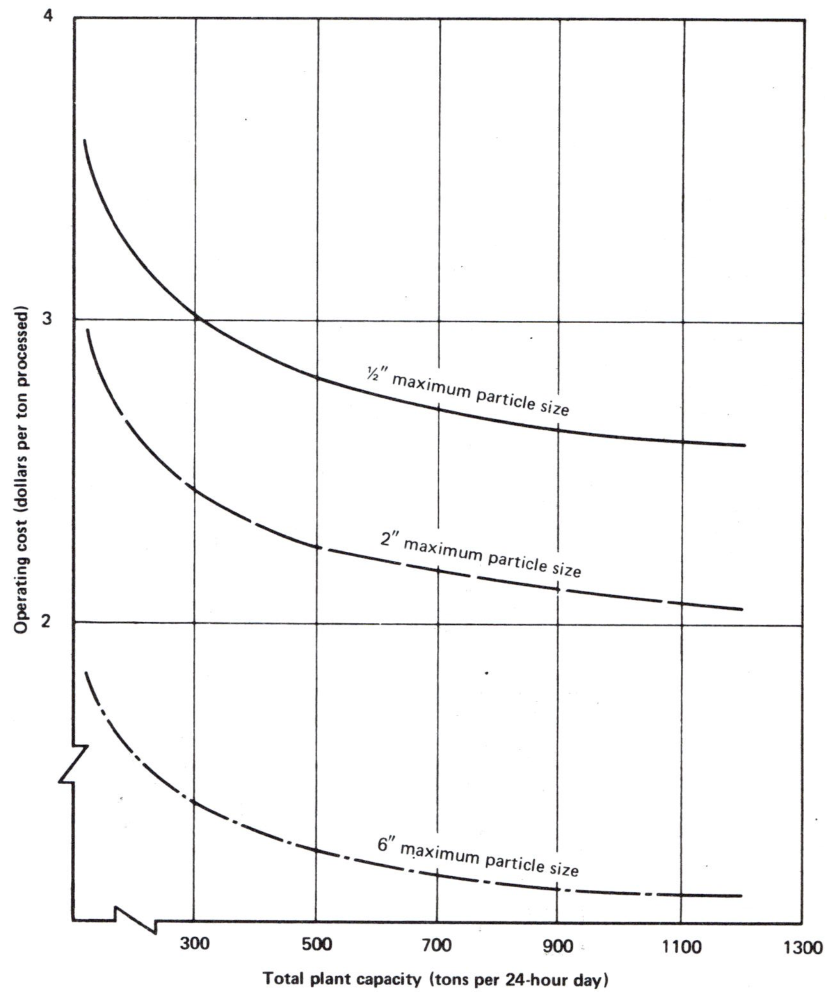

Figure 7.15 Size reduction—total capital and operating costs.

courtesy Arthur D. Little, Inc. [1]

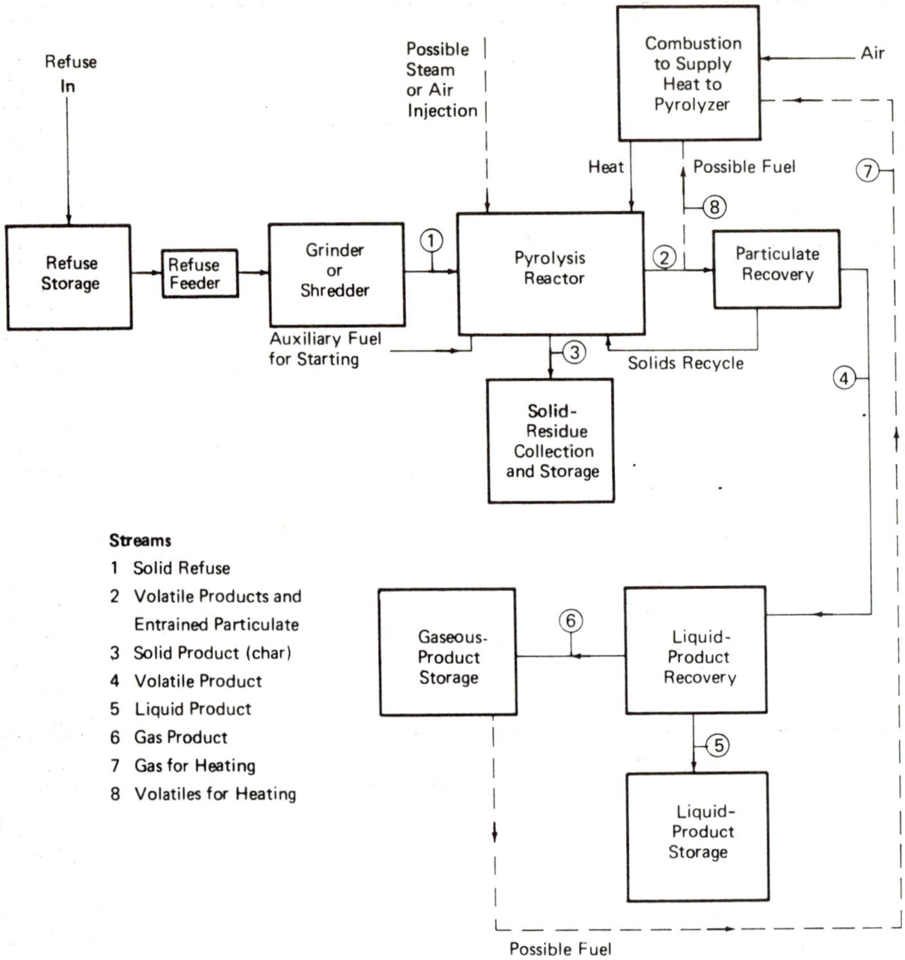

Figure 7.16 Schematic arrangement of refuse-pyrolysis process.

courtesy Arthur D. Little, Inc. [1]

such plants. Comparison with alternative incineration concepts shows dramatically the economic advantages of package systems in small-plant applications. It should be noted that in larger multiples of units than those shown in Table 7.9, the total cost of incineration will be the same (or higher) than the minimum values shown. Thus, in larger plants, the economies of large furnace operation take hold with subsequent improvement in operational economy.

NEW CONCEPTS

The people of the United States have traditionally prided themselves on their innovative exploitation of technology for the material betterment of their lives. In few cases is the need for technological advancement more pressing than in the area of solid-waste disposal. This need, felt through the technical and business communities, has resulted in the postulation of a wide variety of new disposal concepts, many of which are reviewed elsewhere in this book.

A review of new incineration concepts now under development is somewhat disappointing. In general, the concepts are extensions of current practices in incineration or in the pyroprocessing and chemical industries. While this is fortunate from a viewpoint of shortening the development time required, no major new advances covering all five criteria of performance are foreseen.

In reading the newspapers of many of our major cities in connection with the solid-waste-disposal problem, one could conclude that major technological breakthroughs are imminent. This is perhaps, a reflection of either the understandable desire of promotors of new processes to present their work in the best light or of city officials to discover and present to the electorate some miraculous way to provide a "final solution" to the solid-waste problem. The bitter reality is that there are no such magic solutions in the offing. Indeed, the delays in taking action often serve only to intensify the problem. To be sure, some of the new concepts offer promise of improvements in one respect or another, but the imperative is clear: the cities must act decisively and promptly to solve their solid-waste-disposal problems.

Although no giant steps in technology or economics are foreseen, a number of systems now under development offer decided advantages in one or another of the criteria of incinerator performance. Also, in view of the wide interest shown in the popular press, it is appropriate that a brief description be made of the three basic concepts which could be placed in the new-technology category. The stage of development of the processes discussed vary from conceptual design to pilot-plant studies. This evaluation will be made on a generic basis; that is, broad concept categories will be treated and major variations on the basic concepts examined. Minor variations peculiar to specific hardware systems will not be considered in detail.

The categories of the new concepts are as follows:

1. *Slagging incinerators* – involving either combustion of refuse at temperatures above the ash-slagging temperature or fusion of the ash as an additional unit operation following more conventional incineration. Most approaches use a supplemental energy source to reach and maintain slagging temperatures.
2. *Fluidized-bed incinerators* – applying fluidized-bed-combustion technology for the incineration of solid waste. In general, pretreatment of the refuse (shredding, metal removal, etc.) is necessary.
3. *Pyrolysis* – involving the autogenous thermal decomposition of refuse to produce combustible organic gases, liquids, and a high-ash char.

Slagging Incinerators

The principal technical advantages of approaches using slagging techniques are obtaining a maximum reduction in solid-waste volume (in excess of 97 percent) and the assurance of complete combustion or carbonization of all putrescible matter. These advantages are obtained by reducing the refuse residue to a melt, producing a sterile, compact, biologically inert material suitable as a high-quality fill. These advances, unique to slagging systems, are limited to the incinerator performance criterion of residue quality. Other criteria for performance, as discussed above, may be met by the various embodiments of the slagging concept but generally only to a degree equal to or less than that found for modern continuous-feed systems. As a result, slagging concepts may have particular significance and importance only to communities with very limited land for residue disposal or when local situations suggest a revenue-producing market for the residue material (such as use for fill in parts of Florida).

Melting of the incombustible residue can be accomplished either by incineration of the refuse above the residue-slagging temperature or by fusion of the ash in an additional unit after more conventional incineration. In either regard, local temperatures in excess of 2,600-2,800F are required. The minimum temperature which can be used depends upon the metal content and composition of the refuse. To ensure adequate melt fluidity, a temperature of 3,000F may be needed. Some processes utilize limestone or fluorospar additives as a flux to aid fusion and slag fluidity.

In theory, refuse of less than 25-percent moisture could be burned in a special furnace to produce temperatures of 3,000F by operation under or near stoichiometric conditions (to a maximum of 20-percent excess air). In conventional systems, such conditions could not be maintained because of severe operational problems with the grate and the refractory walls.

An energy balance of an incinerator shows that supplementary heat sources will be required to maintain slagging temperatures over almost the entire range of refuse moisture content found in practice. Various approaches have been suggested to supply heat for the slagging process including systems utilizing:

1. Heat recuperation for preheat of combustion air or auxiliary fuel (Dravo/FLK).
2. Coke or natural gas or both in the combustion chamber (Melt-Zit process, American Thermogen Division, Process Plants Corporation).
3. Combustion air preheated in a gas-fired refractory heat exchanger (Torrax).
4. Oxygen-enriched air.
5. A grate-type primary combustion chamber for partial burnout and carbonization of combustibles followed by a coke-fired cupola-type furnace for slagging of the residue from the grate (Ferro-Tech).
6. Combustion air preheated by recuperation operating on refuse from which most metal and glass have been removed to minimize the required temperature for slag fluidity (SIRA).

The above processes differ in the sequence of refuse combustion and slagging steps and in the method of supplying the supplemental energy. They are in various stages of development from full-scale demonstration units to simple process concepts. They all share however, the following characteristics.

1. A volume reduction in excess of 95 percent can be achieved (reducing 1 ton of refuse with a volume of approximately 150 cubic feet to a residual volume of only 3.75 cubic feet).
2. All putrescible material is destroyed in producing the slag.
3. Solidification of the slag can be performed rapidly in a quench tank to produce porous coarse granules, perhaps with the separation of metals. Slow cooling may enable production of a stronger, more useful and valuable aggregate material.
4. Most of the processes utilize less combustion air than found in conventional furnaces. Thus, per pound of refuse burned, the weight of gas passed to the flue-gas conditioning system and air-pollution-control device is less for most slagging incinerators. This reduction should effect economies in the cost of the gas-handling system for the incinerator.
5. The higher temperatures required for slagging may result in substantially more nitric-oxide formation than is generated in normal incineration.
6. Many of the systems are characterized by local high gas velocities which tend to entrain particulates, and thus more efficient particulate-removal devices are necessary to reach the same emission level as conventional incinerators.
7. Supplementary heat (or heat-recovery equipment) is required for almost all concepts resulting in a significant additional operating cost (or amortization expense) above that of conventional incineration.

8. The addition of ash fusion to the incineration process complicates operation. This will make start-up more difficult and may require better trained and higher paid labor. In general, trouble-free operations of slagging processes requires that temperatures remain in a range bounded on the low side by slag-fluidity requirements and on the high side by the limitations imposed by the design and materials of construction. In view of the severe problems caused when these limits are exceeded, it is clear that better efforts in employee motivation or rewards may be needed to avoid high operating costs or frequent outages and malperformance of the plants due to operator inattention.

It is expected that investment costs for slagging incineration systems should not vary significantly beyond the normal spread found in the costs for conventional grate systems. This will be the case despite savings in residue-handling and air-pollution-control equipment associated with the unusual operating characteristics of the incinerator.

In the opinion of the authors, none of the incineration systems utilizing the slagging principle have been developed to the point where any substantial degree of confidence can be had that the operation will be reasonably trouble-free and cost-competitive.[7] It should be expected, therefore, that the initial installations of all of these concepts will be carried out under a federal or state demonstration grant where the technical risk can be at least partially offset by decreased financial obligations. The unique advantages of slagging systems, however, justify such demonstration testing and, undoubtedly, this concept will play an important position in the incineration practice of the late 70's and 80's.

Fluid-bed Incineration

The petroleum and petrochemical industries have established fluidized-bed technology as a powerful tool for performing gas-phase and liquid-phase reactions, especially the former, wherein a number of process variables and controls can be optimized. The high heat-transfer and reaction rates which can be achieved in such fluid beds has suggested their application for the incineration of refuse to achieve good burnout of residue and combustible air pollutants. Although no practical demonstrations of this approach in refuse incineration have yet been made, sufficient interest exists to deal with it here.

Operating Principles

When a gas such as air is passed vertically uniformly through a bed of small solid particles, a velocity is found where the frictional drag on the solid particles lifts the whole bed upward. The bed solids then move in a more-or-less random fashion, and, as a mass, behave very much like a boiling liquid: they become "fluidized."

In the fluidized state, solid particles and gas are in continuous and intimate contact. Low-energy collisions of the particles with each other and with the vessel walls or other surfaces cause a scouring action preventing the occurrence of stagnant layers around these surfaces. Because the solid particles are in continuous contact with the gas phase, are well mixed, and have a large total surface area per unit volume of solids, it is not surprising to find high rates of heat and mass transfer occurring.

All commercial fluid-bed-incineration applications are limited to destruction of relatively homogeneous liquids, slurries, or semisolid mixtures such as dewatered sewage sludges. All of these applications are strikingly similar in operation. They are used to thermally destroy waste

[7]Too little information is available on the economics and performance of the Dravo-FLK process now in operation in Wolfsburg, Germany to permit an assessment of its applicability to U.S. conditions. Insufficient information was available at time of press to provide a description of the Union Carbide Oxygen Refuse Converter. This unit is designed to slag the residue and to produce a low-Btu gas as a by-product.

which contains non-combustible solids that can form either:

1. bed particles which could be removed by overflowing the bed; or
2. solids so finely divided that they would be carried out of the fluidized bed by the flowing gases.

Consequently, the major process variations found between the different suppliers are in the feed methods and mechanisms used for introducing waste into the fluidized beds and the type of equipment used to remove excess solids from the offgas streams.

Refuse Incineration by Fluidized Bed

The use of fluidized-bed techniques for incineration of homogeneous wastes such as oil and sewage sludge is well documented in the United States and German technical literature. Fluidized-bed incineration of municipal refuse is being studied on a pilot-plant scale at West Virginia University. Feasibility studies for high-pressure fluidized-bed incineration of municipal refuse have been made for a system which recovers energy through the generation of electricity. Test evaluations of critical components are scheduled for this system but, at the present time, there are no fluidized-bed incinerators operating on municipal refuse.

Review of the demonstrated behavior of fluid beds in the context of incinerator applications suggests the following characteristics.

1. *Simplicity of construction* – the incinerator consists of a vertical cylindrical chamber with no moving internal parts. High volumetric heat-release rates can be achieved.
2. *Complete combustion at relatively low temperatures* – complete combustion of low-heating-value wastes is possible at about 1,600F. As a result, combustible pollutant emissions should be small.
3. *Low NO_x emissions* – the low operating temperatures and absence of local high-temperature combustion zones or hot spots should result in the production of negligible amounts of nitrogen oxides.
4. *Low flue-gas volumetric rates* – complete combustion at excess-air rates as low as 10 percent results in a very significant reduction in the amount of gas to be cleaned relative to conventional incinerators operating at excess-air rates of 200 percent. In order to operate at these low excess-air levels and maintain temperatures in the range of 1,600F, however, some water addition may be needed, thus increasing the flue-gas flow rate from the "theoretical" optimum.
5. *Operating stability* – the large thermal capacity or inertia of the bed tends to damp out moderate temperature or other operating fluctuations due to short-term variations in feed characteristics.
6. *Ease and efficiency of intermittent operation* – the excellent heat-retention capabilities of a static bed of solids (with adequate wall insulation) enables the fluidized-bed incinerator to be shut down for periods of time without the temperature dropping below a satisfactory level for immediate startup. If substantial cooling has occurred, only a short waiting time is necessary prior to beginning incineration.
7. *Feed preparation required* – considerable feed preparation will undoubtedly be required to assure retention of particles in the beds to complete the combustion of refuse and to permit the removal of incombustibles.
8. *High particulate furnace emissions* – the high gas velocities characteristic of this device will result in high solids loading in the flue gas. Thus, highly efficient particulate-removal

equipment is mandatory to achieve emissions comparable even to those from conventional incinerators. The higher-efficiency requirement could easily offset the investment reductions due to low gas volume if comparable particle-size-distribution particulate is produced in the fluid bed.

9. *Unit-size limitation* – the maximum single unit size for fluidized-bed refuse incinerators operating near atmospheric pressure is estimated to be 50-60 TPD; the size limitation corresponds to the maximum cross-section consistent with good bed distribution. This size limitation is very unfavorable to good operational and investment economics. Higher operating pressures would increase the throughput for a unit of given volume but would incur problems in refuse feeding and residue removal.
10. *High power consumption* – power consumption will be very high relative to conventional incineration systems due to the power requirements for fluidizing-gas blowers and refuse preparation.
11. The cost per TPD for fluid-bed incinerators will be almost prohibitively high. Costs for the basic fluid bed unit will be expected to approximate $6,500 per TPD. Adding peripheral equipment, refuse preparation and so forth results in a total cost, even in large plants, approaching $12,000 per TPD. Based on anticipated use of approximately 160 kilowatt hours per ton of refuse and the necessity for shredders and other equipment to pretreat the refuse it is estimated that even in a plant as large as 600 TPD, the operating cost for three-shift operation would be in excess of $10 per ton. Although at the penalty of some additional capital investment, heat recovery could be utilized to generate steam to drive the combustion-air blowers, but incineration costs in large plants would still be in excess of $7.50 per ton. This is at least twice as high as for other incineration concepts which are evaluated on the same basis.

In view of the large number of technical and economic problems associated with the fluid-bed-incineration concept, it is not envisioned that fluid-bed technology will find acceptable applications for municipal refuse incineration without a major developmental effort. A pilot-scale unit is currently being utilized by the Combustion Power Company, under contract to the Bureau of Solid Waste Management, to resolve some of the problems associated with fluidized-bed incinerators and to evaluate the potential of fluidized-bed incinerators as a source of energy for gas turbines.

Pyrolysis

The pyrolysis of carbonaceous solids is a destructive-distillation process effected by the application of heat to the solids in the absence of air. The solid decomposes into gaseous, liquid and solid products.

In general, the gaseous products contain carbon dioxide, water vapor and a mixture of combustible hydrocarbons; the liquid is a mixture of water, oils and tar; and the solid is a carbon-rich, high-ash residue. The quality and quantity of products obtained depends upon many factors, some of the most important being the type of carbonaceous solid employed, the ultimate temperature attained, the heating rate, and the type of equipment used. Within certain limits, manipulation of these variables offers a mechanism for controlling the quantity and yield of given products.

Although the details of pyrolysis behavior can be discussed only in the context of a given solid material and pyrolysis process, certain basic principles obtain. Both chemical and physical changes occur in solids during pyrolysis. The most important physical change in materials such as bituminous coal and some plastics is a softening effect followed by a resolidification resulting in a solid residue. Other solids, including many plastics found in refuse, simply melt and do not resolidify unless they are cooled. Cellulosic materials such as wood and paper increase in porosity and swell a certain amount as the volatile components are evolved. These physical changes are important because of their effect on the handling of material in the pyrolysis equipment and as they affect the reactivity of the solid residue.

The chemical changes occurring during pyrolysis involve reactions that decompose the original material into products. Primary products are evolved directly from the complex organic molecules that are typical of organic solids. As heating proceeds, the primary products further decompose or crack, forming secondary products, usually with the deposition of solid carbon. The composition of the final products depends on the heating rate and the ultimate temperature attained. As the ultimate temperature increases, the secondary products become simpler in structure and more hydrogen-rich in composition. At the same time, the solid residue becomes more like graphitic carbon in both chemical composition and physical structure.

The temperature at which the rate of evolution of pyrolysis products becomes significant depends upon the solids being heated. In the case of wood and paper, it is known that a substantial amount of loss of mass occurs in pyrolysis at temperatures as low as 400F, whereas temperatures approaching 700F are required for a significant mass loss from coal. Behavior of solid refuse would be expected to be similar to that of wood and paper.

Whether the overall pyrolysis process of a given solid requires or yields heat depends upon the ultimate temperature attained. In the case of most materials, the process is endothermic (requiring heat addition) at lower temperatures and exothermic (yielding heat) at higher temperatures. Thus, as the material is heated, energy is first absorbed and then released. If the ultimate temperature is high enough, the net energy transfer will be away from the material.

Current Status of Refuse-pyrolysis Technology

The process of pyrolysis as applied to municipal refuse is a relatively new concept, the technology of which is very limited. The knowledge of this process that does exist has been generated mainly in laboratory investigations, but some data are beginning to appear from full-scale process demonstrations. The following discussion outlines the present knowledge of the process in the context of a general idealized flow diagram of the refuse-pyrolysis concept.

Various arrangements of equipment can be visualized for pyrolyzing refuse. In general, the process will employ a storage bin, a feeder, a grinder or shredder (for sizing rather than for size reduction), a pyrolysis reactor fired with part or all of the volatile pyrolysis product, a particulate-recovery unit, and units for collecting and storing the solid, liquid and gaseous products. A general process diagram is shown in Figure 7.16.

Different designs for the reactor, the heart of the unit, can be envisioned. One that has been used on a full-scale process is an air-tight cylindrical retort that rotates slowly on a slight incline and is fired externally with pyrolysis gas. The refuse enters and leaves respectively at the higher and lower ends of the retort. The main design used in laboratory tests to date has been a batch retort heated indirectly with electricity. Laboratory investigations have considered the use of a fluidized-bed pyrolyzer, but such a unit has not yet been demonstrated as suitable for solid-refuse pyrolysis applications.

Shredding equipment is included in Figure 16 because the times required to pyrolyze large pieces of refuse appear to be prohibitively long. Because heat-transfer rates apparently control pyrolysis rates, it appears that the refuse should be reduced to sizes less than about four inches for plates and less than about two inches for chunks if the required residence time in the pyrolyzer is to be less than 45 minutes to one hour.

It is worthy to note that the Torrax slagging incinerator incorporates a pyrolysis step in the process accomplished by partial combustion of the char with preheated air in a shaft-furnace.

Pyrolysis Products

The organic liquid product of refuse pyrolysis contains an incompletely described mixture of alcohols, oils and tars known as pyroligneous acids. It is known that the liquid fraction distilled from typical refuse is a highly complex mixture of many organic compounds containing 70-80 percent water, the remainder being various alcohols, ketones, tars, acetic acid and other organics. Methanol; 2-ethyl, 1-propanol; 1-pentanol; 3-pentanol; 1, 3-propanediol; 1-hexanol; and acetic acid have been separated and tentatively identified. Based on present market values, it would appear that although some of these materials have relatively high commercial value and developed markets, separation and purification processes would be prohibitively expensive. As a consequence, it is likely that the liquid product would be considered only for its fuel value, where it is expected that a characteristic heating value might be between 1,000 and 2,000 Btu per pound of combustible refuse.

The yield, as well as the calorific value, of the gaseous products of refuse pyrolysis is highly variable. The mean gas yield at around 1,500F is about 25 percent by weight of the air-dried, ash-free refuse. It would appear, based on limited data on a synthetic refuse, that the calorific value might approximate 300-350 Btu per cubic foot at a yield of 2-3 cubic feet per pound of refuse combustibles.

The solid product or char resulting from refuse pyrolysis is an impure carbon (excluding the ash), very similar in proximate analysis to coal. The yield of char, expressed on an ash-free basis, is between 17 and 32 percent of the air-dried refuse weight; the variation results from different refuse compositions. The mean yield is about 17-25 percent by weight of the air-dried ash-free refuse, decreasing with both increasing heating rate and increasing pyrolysis temperature. The calorific value of the char is about 11,000-12,000 Btu per pound of air dried, ash-free char and appears to decrease slowly as pyrolysis temperature increases. The ash content of the char has not been well defined by experiment but probably varies between 50 and 60 percent. Since this is a rather high ash content for a fuel, it will be desirable to remove some of the ash-forming materials such as glass, metal and other inert objects before or after pyrolysis in order to improve char quality.

Evaluation of the Refuse-pyrolysis Process

The pyrolysis of refuse is new: evaluating its advantages and disadvantages is consequently somewhat difficult, owing to the lack of information. This is especially true with respect to economics. However, some preliminary speculation based on the general features of the process is useful.

Four main advantages of the process are often given.

1. The process should facilitate the control of air pollution.
2. Useful products might be produced from the refuse.
3. The reduction in the volume of the refuse and the production of a sterile residue material would extend the life and improve the quality of sanitary landfills relative to untreated refuse, and
4. The process is self-sustaining with respect to energy.

The basis for the first advantage is that the process is completely contained and it produces relatively low gas volumes per ton of refuse. Since the process is contained, any pollutant generated therein can be theoretically collected. The relatively low temperatures and low oxygen concentration within the pyrolyzer suggest that the generation of oxides of nitrogen will be low. However, if the volatiles or the char are used to supply the heat for pyrolysis, such secondary combustion devices will presumably involve temperatures and oxygen concentrations which will result in nitrogen-oxides formation.

The relatively low gas velocities and volumetric rates of generation of pyrolysis gases should result in a very low level of particulate carryover. Removal of particulate from these gases can probably more easily be performed after combustion as condensible components would probably foul particulate-removal systems. The gas volumes and required efficiencies would be very much lower than particulate removal for conventional systems.

A second, and possibly more important, consideration regarding improved air-pollution performance is based on the physical transformations effected in the pyrolyzer. Refuse is a heterogeneous, variable, unstorable, bulky, hard-to-handle fuel requiring special combustion devices. Pyrolysis products, with minor blending, are homogeneous, consistent, storable, compact, easy-to-handle fuels, probably readily combustible in conventional furnaces with conventional burners. The effects of such a transformation on the ability to burn these materials with a minimum of air pollution and to advantage are obvious and most encouraging.

The advantage of producing useful products from refuse depends on economic factors that are not analyzed here. Such an analysis is difficult not only because the exact nature of the product is not well understood, but also it is not yet clear how much of the liquid and gas will be available after supplying process energy requirement. It is clear that the unit Btu content of the results of the pyrolysis process are not high in comparison to ordinary fossil fuels and thus only moderate transportation charges to the place of ultimate consumption can be accommodated. Such products, however, could provide a desirable couple between a refuse-disposal system and an electric utility.

It has been suggested that the char produced by pyrolysis could find application (before burning) for tertiary treatment of sewage-treatment-plant effluent. Although refuse could provide only a partial source of this char for a given population, it would be available at a nominal cost.

The volume reduction effected by pyrolysis is quite modest compared with that achieved by conventional incinerators. Volume reduction of the combustible portion of refuse is about 50 percent. Therefore, pyrolysis will extend the life of landfills by less than a factor of two. As a result, the effectiveness of pyrolysis-based concepts in volume reduction is probably unacceptable.

The Lantz converter, a refuse-pyrolysis system, developed by Pan American Resources (PAR) of Albuquerque, New Mexico, is comprised of a sealed, air-tight retort cylinder inside a heavy insulated jacket. The retort is gas-fired, usually about 20-feet long, and revolves slowly on a slight incline from infeed to outfeed. Refuse is admitted through a seal that intermittently opens in such a way that the entry of air is minimized. The retort pyrolyzes refuse at temperatures ranging from 900F to 1,500F depending on the type of refuse.

A full-scale Lantz converter was installed and operated by PAR at a Ford Motor Company plant in San Jose, California, and began operation in 1968. Tests run on the system reportedly designed to handle 40-50 TPD of plant refuse (paper, wood and other packaging materials, print sludge and some plastics) indicated that the pyrolysis scheme would work successfully on this refuse. Operation of the system was discontinued, however.

Economics

The capital costs for pyrolysis furnaces should be similar to those of conventional incineration furnaces. Buildings, feed, and char handling should also be equivalent to the comparable components of conventional systems. Differences will arise in the feed preparation and air-pollution-control equipment costs.

It is not expected that any useful or salable chemical products (in contrast to fuel products) can be economically separated from the liquid, gaseous or solid products of refuse pyrolysis. Instead, a portion of the products would be burned to supply the energy required for the pyrolysis reaction. Any excess could be flared, perhaps used to supply plant steam, or if available in sufficient supply, marketed as a low-sulfur fuel.

In view of the large number of technical problems and economic questions surrounding the pyrolysis process, it is difficult to determine the relative importance of this approach to solid-waste disposal in the future. Because of several potential advantages of the system with respect to air pollution and perhaps some cost savings, the concept certainly is worthy of further attention.

SUMMARY OF INCINERATION ECONOMICS

Too often, economic constraints are applied to technical decision making regarding only part of the proposed design or operating practices for incinerators. That is not to fly in the face of the need for economy in public services. On the contrary, it is false economy that is a concern. Municipal waste disposal is reputed to fall only behind education and highways on the scale of municipal expenditures. It is reasonable, therefore, in this important area of refuse waste disposal by incineration, to consider carefully the impact of cost-cutting on system performance and operating costs.

Need for High Priority in Furnace Design

In reviewing Figures 7.9 and 7.10, it can be seen that the basic refuse-processing unit, the furnace, comprises only about half of the total plant cost. Yet, in the often-occurring case where bonds are issued against incinerator construction long after the original estimates were prepared, when inflation has escalated the cost, the furnace is too often the first target of cost-cutting (perhaps only after the air-pollution-control device). It is not reasonable, from a technical or performance standpoint, to strike first at the very heart of the process. Similarly, air-pollution control, though often the major source of complaints concerning incinerators, constitutes only 10 percent of the investment cost in larger plants. If a reduction of 5 percent in the total plant cost is taken from this segment of the plant, however, performance drops from the 90+ percent particulate-control-efficiency level to 40 percent (a six-fold increase in emission rate). This is an unacceptable "trade-off."

Heat Recovery Usually Difficult to Justify

Communities considering waterwall systems should review carefully the potential reimbursements from selling steam. The economics shown above suggest very strongly that steam credits are necessary to justify the incremental costs of such systems, particularly in the case where suspension burning is proposed. Therefore, although heat recovery has the emotional and political appeal of extracting a useful commodity from refuse, the decision to employ this design concept can bring on economic hardships which threaten the remaining portions of the process.

The concept of on-site electrical power generation in conjunction with conventional incineration does not appear economical in small plants at present fuel costs in the United States. This is primarily due to the high initial cost of electrical-generation and steam-raising equipment and the high cost of additional skilled labor at the small scale associated with refuse incineration. The possibility of combined firing or the utilization of steam generated in an incinerator in a nearby power plant comes closer to meeting acceptable economic criteria for electrical generation from refuse incineration.

Building Costs Can be Reduced

The costs of the buildings and grounds for incinerators typically constitute almost half of the investment. Simplifying designs through the use of outdoor construction methods regularly used in the chemical and metallurgical industries should be explored. While it is true that such plants of themselves may appear unsightly, the careful use of screening can avoid the "manufacturing-plant" appearance and can serve to enhance the aesthetic appeal of the entire facility.

Design Stage Must Not be Starved of Funds

The cost of an incinerator is high, and its lifetime usually spans at least 20 and often 30 years. It is most appropriate that a comprehensive design effort be carried out. Too often funding limitations give the consulting engineer little latitude in the detail of his process-unit design effort and cause him to emphasize the mechanical process of producing drawings. Such a misplaced priority is unrealistic and perpetuates design deficiencies which could be perceived and solved but for the press of funds. The comprehensive systems-engineering approach to incinerator design is quite justified by the cost of the device and can be repaid many-fold in reduced operating costs and improved performance.

Batch Feeding and Shredding

The small municipality considering a new incinerator should explore the potentials of advanced batch-feed systems. Although continuous-feed furnaces offer the optimum in performance, a well-designed batch-feed plant, equipped with proper feeding devices and air-pollution-control devices, makes economic sense for small installations and can perform satisfactorily. For the very small plants, package incineration systems show significant advantages.

All incineration concepts requiring the use of refuse shredding (such as suspension-burning boilers or other concepts) should be recognized as incurring substantial economic handicaps. The cost of shredding can be shown to add at least $2 and as much as $4-$5 per ton to the total incineration costs. In addition, processes which depend upon the shredders (typically a high-maintenance, high-outage device) can suffer unreliabilities beyond those inherent in the incineration system itself.

CONCLUDING COMMENTS

Incineration provides a means for alleviating waste-disposal problems in regions where space for landfills is not readily accessible. Many of the problems associated with current incineration practice in the U.S. stem from a lack of appreciation for the fact that an efficient incinerator is a complex chemical plant in which solid, liquid, and gas streams must be processed.

Municipalities often do not have the background or technical assistance required for drawing specifications that are reasonable and consistent and they are constrained in their selection of design by financial pressures which favor units with low first costs over more-expensive better-designed units that will yield the specified performance at a minimum overall cost. Furthermore, the municipalities are often unwilling to assign salaries and recognition to incinerator personnel commensurate with the skills required by their positions. These problems are compounded by the wide variation in design, in cited disposal costs, and in potential unit performance in the bids submitted to municipalities, and by the absence of any party that will accept responsibility for the total system performance.

Some of these problems could be alleviated by a regional approach to waste disposal. Benefits that would result include the following.

1. Reduction in capital and operation costs from scale-up, (see Figures 7.9 to 7.13).
2. The safeguard provided by multiple facilities. This would permit the design of larger, more economical furnaces in a single plant.
3. Attraction of competent technical and managerial personnel and of long-range financing.
4. The increased potential for by-product recovery of scrap metal, steam and residue.

A model for such regional treatment is that of the urban centers in Europe. One example is Traitement Industriel des Résidus Urbains [19, 27], now a subsidiary of Electricité de France, which has been treating Paris' waste on a regional basis for some 50 years. Regional treatment of solid wastes has been considered by several communities in the U.S. but with the occasional exception [26] the plans have been thwarted by the difficulty of finding host communities for the incinerator units.

A number of new incinerator designs are under development and provide promise of improved operation and, for slagging, pyrolysis, and energy-recovery units, of revenue from by-product sales partially offsetting disposal costs. It is difficult to assess when these units will become operational but their development should be encouraged by the demonstration-grant program of the Bureau of Solid Waste Management.

REFERENCES

1. W. R. Niessen, et al "Systems Study of Air Pollution from Municipal Incineration," *Final Report* under Contract CPA-22-69-23 to the National Air Pollution Control Administration, Arthur D. Little, Inc., Cambridge, Massachusetts, March, 1970.
2. E. R. Kaiser, "Chemical Analysis of Refuse Components," Proceedings 1966 National Incinerator Conference, American Society of Mechanical Engineers, New York, May, 1966.
3. E. R. Kaiser, C. D. Zeit, and J. B. McCoffery, "Municipal Incinerator Refuse and Residues," Proceedings of 1968 National Incinerator Conference, American Society of Mechanical Engineers, New York, May, 1968.
4. Proceedings of 1964, 1966, 1968 and 1970 Incinerator Conferences, American Society of Mechanical Engineers, New York, 1964-1970.
5. R. C. Corey, Editor, "Principles and Practices of Incineration," Wiley-Interscience, New York, 1969.
6. J. DeMarco, et al "Incinerator Guidelines – 1969" Public Health Service Publication No. 2012, Bureau of Solid Waste Management, 1969.
7. "I.I.A. Incinerator Standards," Incinerator Institute of America, May, 1966.
8. S. H. Chansky, "Analysis of Refractory inside Surface Temperature versus Refractory Thermal Conductivity and External Heat Removal Systems," paper presented at American Society of Mechanical Engineers, Incinerator Division, New York, September, 1969.
9. Personal communication. D. G. Wilson, July, 1968.
10. Brennst-Warme-Kraft, *17*, No. 8, (1965).
11. Atlas Systems, Storage Bin and Reclaiming Systems, Brochure, Spokane, Washington.

12. A. H. Rose, et al "Air-pollution Effects of Incinerator Firing Practices and Combustion-air Distribution," *Journal of the A.P.C.A., 8*, 297-301 (1959).
13. R. L. Sternburg, et al "Effects of Design and Fuel Moisture on Incinerator Effluents," *Journal of the A.P.C.A., 10,* 114-20 (1966).
14. R. L. Sternburg, et al "Field Evaluation of Combustion-air Effects on Atmospheric Emission from Municipal Incinerators," *Journal of the A.P.C.A., 12,* 83-9, (1962).
15. J. A. Fife, and R. H. Boyer, "What Price Incinerator Air Pollution Control," Proceedings of the 1966 National Incinerator Conference, A.S.M.E., New York, May, 1966.
16. Personal Communication. J. A. Fife, July, 1968.
17. F. Nowak, "Considerations in the Construction of Large Refuse Incinerators," Proceedings of 1970 National Incinerator Conference, A.S.M.E., New York, 1970.
18. F. Nowak, "Corrosion Problems in Incinerators," *Combustion, 40,* 32-40, (1968).
19. H. Rousseau, "La Récuperation de Chaleur dans les Usines d'Incinération de la Région Parisienne," 3eme Congrès International, International Research Group on Refuse Disposal, Trente, Italy, May, 1965.
20. E. W. Kiss and C. O. Velzy, "Oceanside Refuse-disposal Plant Illustrates Boiler Application for Improved Economy," Presented at 56th Annual Meeting, National District Heat Association, Williamsburg, Virginia, since 1965.
21. A. B. Walker and F. W. Schmitz, "Characteristics of Furnace Emissions from Large, Mechanically Stoked Municipal Incinerators," Proceedings of 1964 National Incineration Conference, A.S.M.E., New York, (1954).
22. C. J. Stairmand, "Removal of Grit, Dust and Fume from Exhaust Gases from Chemical Engineering Processes," The Chemical Engineer (Great Britain), *194* (12), 310-26, (1965).
23. C. Rampacek, "Reclaiming and Recycling Metals and Minerals Found in Municipal Incinerator Residues." Proceedings of Symposium on Mineral Waste Utilization, pp. 124-131, March, 27-8, 1968.
24. G. Stabenow, "Performance and Design Data for Large European Refuse Incinerators with Heat Recovery," Proceedings of 1968 National Incinerator Conference, pp. 278-286, A.S.M.E., New York, (1964).
25. A. B. Walker, "Electrostatic Fly-ash Precipitation from Municipal Incinerators; a Pilot-plant Study," Proceedings 1964 National Incinerator Conference, A.S.M.E., New York (1964).
26. Anonymous. "A Regional Approach to Refuse Disposal," *American City,* June, 1964.
27. H. Rousseau, "The Large Plants for Incineration of Domestic Refuse in the Paris Metropolitan Area," Proceedings, 1968 National Incinerator Conference, pp. 225-231, A.S.M.E., New York, 1968.

CHAPTER 8

RECLAMATION FROM MUNICIPAL SOLID WASTE

The incentives promoting, and the barriers restricting, greater reclamation of valuable components from mixed municipal refuse are briefly reviewed in this chapter. Existing methods of sorting refuse into its useful and other components are described, and some of the current approaches towards mechanizing this process are outlined. It is concluded that the prospects for successful and economic reclamation of some of the major components of domestic refuse in many circumstances are bright.

INTRODUCTION

In the old days Grandma saved the string and wrapping from Christmas parcels, and we bundled up our old newspapers for our public-works departments or our churches to collect and to sell to paper merchants. The reasons why these wholly beneficial actions do not generally take place at present seem obvious but are worth restating.

The principal reason is that we have achieved such advances in mechanization of production that the number of man-hours needed to supply us with our consumer goods is now a very small percentage of what it was in the past. It may, and probably does, require more man-seconds of labor for a householder to bundle up the week's newspapers, and for these newspapers to be taken through the many stages of handling, baling, transporting, more handling, and repulping, than it does for trees to be cut, debarked, transported, chipped and pulped to make fresh newsprint. Because of the distorted economics and the political complications connected with the recovery of old newspapers we have in the past devoted all our inventiveness and energies on mechanizing the process for production of newsprint from trees, and almost no mechanization has been introduced to collection and handling of so-called "secondary fibers."

The second reason for this state of affairs is that we have seemed to have inexhaustible resources on the one hand, and large tracts where waste materials could be dumped on the other. Quite suddenly the cries of the conservationists that we were deluding ourselves have been heard. By a policy of consume and discard we are destroying irreplaceable natural resources at both ends of the overall process: denuding a hillside or filling a bog are both death sentences to the fragile ecologies of these regions. Further, regardless of our feelings, or lack of them, for the sanctity of the various forms of life in these areas, the breaking of biological chains will and has seriously affected our human existence in many ways, only a few of which are yet understood.

Accordingly the need for better methods of recycling our resources is apparent. The emphasis, when this goal is presented to the public, often seems misplaced. To many people the problem of reclamation is simply one of finding ways to reuse discarded materials. In fact, there are excellent uses available at the present time for most discarded materials. However, for these uses to be economic in the market place, two conditions must be met.

1. The materials must be of sufficiently high purity.
2. They must be at the right place at the right time.

THE ECONOMICS OF PURITY

Conceptually one may draw smooth curves relating value per ton to the percentage of contaminants for each material and contaminant, (Figure 8.1). With some materials, the amount of contaminant which can convert an asset into a liability is very small. Municipal sewage, although about 99.9 percent pure drinking water, has a negative value for almost any known purpose. Scrap steel becomes a liability when the percentage of copper rises above 0.02 percent (however, if scrap steel having more than 0.02 percent copper were segregated – a form of purity – it would be useful in making stainless and other alloy steels).

In practice, the value of reclaimed materials is classified into discrete categories with maximum percentages of specified contaminants, so that the stepped line of Figure 8.1 more nearly represents the true situation. Publications such as the Waste Trade Journal [1], Fiber Market News, [2] or the standards of the Paper-Stock Institute of America [3], list classifications of many 'secondary' materials (as reused materials are called) and, in the Waste Trade Journal and Fiber Market News, current prices are listed.

These standards are not merely ideals. They are rigidly enforced. If a paper mill finds one foam-plastic drinking cup squashed into a bale of paper which has been sold to them as No. 1 News, the vendor will be ordered to remove the entire shipment and his payment will be stopped.

It is necessary to keep in mind the question of contamination when considering the other factor in reclamation – how to get the material to the right place at the right time.

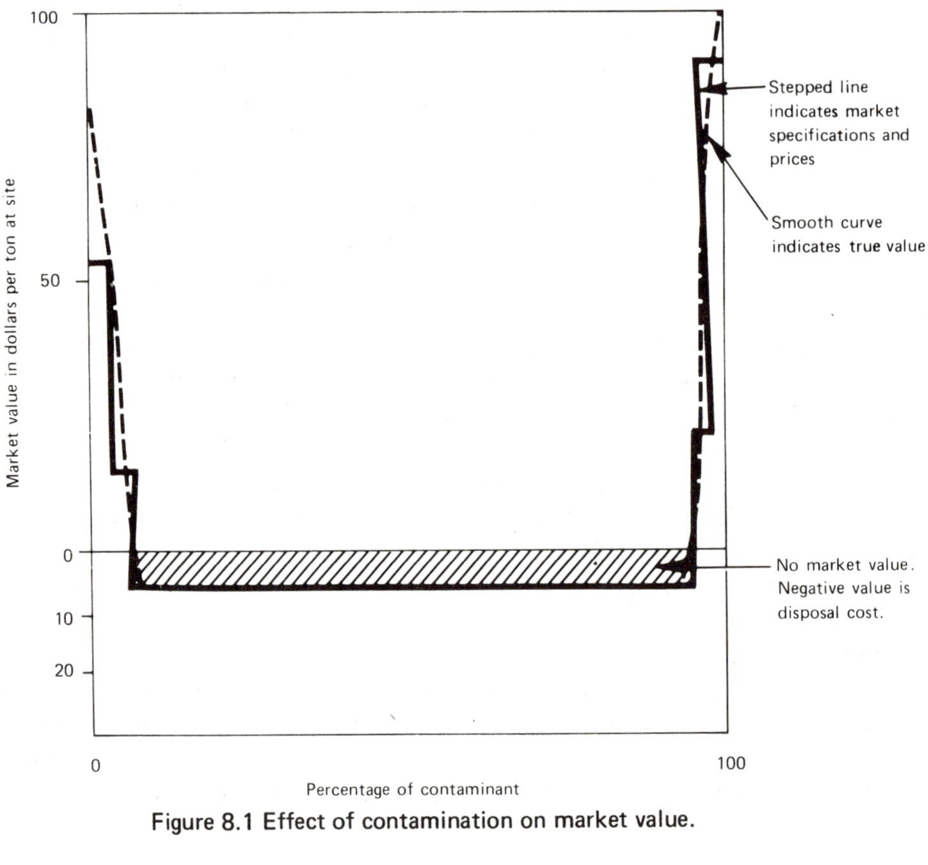

Figure 8.1 Effect of contamination on market value.

SEPARATION AND TRANSPORTATION

Most waste tends to become contaminated – its entropy increases, in thermodynamic terms, and it is then necessary to work to decrease its entropy again. However, there is large category of waste which, with little care, can remain uncontaminated. This is the waste produced in the plant or factory, the so-called "home scrap." Examples are newspaper over-runs; envelope clippings; plastic trimmings from a molding plant; metal turnings from a machine shop (if segregated by material); and risers and poor castings in a foundry. In these cases reclamation has a double advantage, since the path to the point at which the scrap can be reprocessed is usually a short one.

Reclamation of "home scrap" is therefore easy; it has always been practiced for most materials, and the inclusion of home scrap in reports of quantities of materials recycled can be deceiving unless the origin of the materials is made clear. For instance, it is claimed that 100,000 tons of paper are reclaimed every week in the U.S.. However, a large proportion of this amount is home scrap from newspaper overruns and clippings from trimmed operations; the quantity reclaimed from domestic and other use is small and indefinite.

It is reclamation from the huge quantities of domestic and light commercial wastes which is the goal of many people and groups. The technical problems involved will be discussed later. These problems are small compared with the "people" problems – the institutional, legal or traditional blocks to reclamation. One major problem is the wild fluctuations in the market price of many secondary materials, particularly of paper, which is the principal component of domestic (and therefore of municipal) trash. At the time of writing (March, 1971) the price paid for clean newspaper by dealers in the Boston metropolitan area has fallen to zero, whereas a year ago \$10 to \$25 per ton could be obtained. No business can base itself on such a market. (Paradoxically, low-grade paper is being bought by a roofing-materials manufacturer, at a time when housing starts are at a low point.)

Another gloomy picture of the prospects for recycling was given by Leonard Stefanelli, president of the Sunset Scavenger Company [4]. The peak year for recycling for Sunset was 1954, when more than \$500,000 was realized from salvage sales. The following year, according to Mr. Stefanelli, America entered the no-deposit, no-return era, and the type of problems that occurred are worth describing because they would not be foreseen by the ordinary citizen or municipality contemplating setting up a salvage plant.

For instance, champagne bottles, formerly worth almost 15 cents each, began to be fitted with plastic stoppers which required precise-contoured necks and were not reused. Sunset had to discard 16,000 bottles.

But Sunset used to store broken white-glass (flint-glass) bottles until about 80 tons had been accumulated, when the material would be loaded on to gondola cars to be shipped to San Francisco or Los Angeles where it would fetch \$20 per ton. Then the price dropped to less than \$3 per ton, which was not enough to pay the freight.

Not all U.S. salvage companies are as pessimistic. The Proler Corporation and others which, through mechanizing the breaking down of scrap automobiles, have brought about such beneficial changes in the appearance of many of our roadsides, is making a respectable profit after paying collectors about \$13 per ton on site. The "catchment area" for a single large automobile-shredding plant, such as Prolerized New England in Everett, Massachusetts, extends over a radius of 100-150 miles resulting in the creation of subsidiary industries for crushing the hulks and hauling them to the plant.

The upswing in prices of low-grade steel scrap (such as that from automobiles and appliances) has been due largely to a technological development – the introduction of numbers of electric-arc furnaces, which can operate on 100-percent scrap. (These furnaces counterbalanced the adverse effect of an earlier technological development in the steel industry, the changeover to basic oxygen furnaces, which could take much less scrap than the older open-hearth furnaces). But much of our reclaimed scrap is shipped overseas to Japan and elsewhere, where the economics of recycling may be a forecast of conditions to come in the U.S.

Likewise the conditions for paper salvage in Britain are illustrated by the title of an article by Frank Flintoff, who can speak with great authority, "No risk of over-production in growing waste-paper market" [5].

Salvage as a money-making proposition on its own presents, therefore, a mixed prospect at present. Some reasons for this state of affairs are proposed below. However, salvage may be highly attractive when it is considered as an alternative method of disposal in high-disposal-cost areas. In metropolitan regions, where the cost of incineration or of trucking to a distant landfill may approach or even exceed $10 per ton, this amount can be taken as an immediate credit when materials are salvaged. It is possible to make a profit even when nothing is paid for salvaged materials, so long as they are removed at no cost. This theme will also be enlarged upon below.

On-site Separation

Some communities have ordinances which require householders and businesses to separate solid waste into up to five categories (paper, cardboard, glass, cans, garbage and the remainders). This approach works in only a few places, and has been abandoned in all large cities because of the obvious unpopularity of such regulations. As has been suggested above, if the householder's time is correctly evaluated, such a procedure may actually be economically wasteful despite the savings in resources (i.e. greater savings would be achieved if the same effort were concentrated in alternative methods).

However, a strong case can be made for on-site separation of newspapers, telephone books, and magazines, and of corrugated-cardboard boxes, especially when they are bundled or baled. In the first place, these items often constitute so large a proportion of the solid-waste output of an average home, in the case of newspapers and magazines, or of a business, in the case of the corrugated containers, that baling them separately saves on the use of containers. And containers, valuable as they are for loose, miscellaneous trash, have to be handled: in the case of the domestic refuse barrel, it has to be taken back into the premises again, whereas a bundle of newspapers is put out and forgotten.

In the second place, a bundle of newspapers or of magazines (preferably they should not be bundled together because each can go to a higher-value end use when separate) is of large-enough density that an expensive compaction truck is not necessary for pickup; any pickup truck or station wagon will do.

Thirdly, and most importantly, picking up newspapers and baled corrugated with pickup trucks is a comparatively quiet procedure. This means that it can take place in the early morning or late evening at times when use of a regular compaction truck and the rattling of ash cans would raise immediate citizen complaint. Waste-paper collections can be made when the roads are free of congestion at relatively high speed. No crew is necessary; the driver can pick up bundles without difficulty. Compared with the average nationwide of about $18 per ton for regular refuse collection, picking up bundled waste paper and newspaper from the roadside cannot average more than $5 per ton. Whereas a federally assisted demonstration program has been underway in Madison, Wisconsin where newspaper is deposited separately in racks slung under the regular municipal collection vehicle, and the program is reckoned a success, it would be expected that considerable savings would result from separate collection, as has been the trend in Europe.

For other materials, on-site separation and collection seem justified only in special circumstances. For instance, a community may have a special aluminum-recycling drive, or a clean-up week in which the municipality, or a contractor, picks up large appliances and other predominantly steel items for delivery direct to a secondary-metals merchant. As a regular weekly procedure, separation of anything except paper does not seem worthwhile.

Central-station Salvage

If on-site separation and collection is not economic for most materials, the alternative methods of conserving resources are some form of prohibition of use, such as the bans being widely considered and applied against nonreturnable bottles and other forms of packaging, or of reclamation from the mixed state. Mixing means contamination, but the materials which suffer most from contamination are fibers such as paper and cardboard, which, as has been claimed, might be picked up separately. However, a way around the contamination problem could be to supply households and businesses with distinctively-colored plastic bags – white for newspaper, yellow for magazines, brown for cardboard, perhaps – and to use a collection truck which did not break the bags during the compaction process. Whether or not such a procedure would bring benefits (from decreased contamination) greater than the cost and effort involved can be found only by a trial.

Using bags in this way is one form of "coding" materials to facilitate the sorting process. It has frequently been suggested that all materials might be coded in production to enable sorting and reclamation to be carried out much more easily. Three possibilities mentioned are the use of fluorescent dyes, of magnetic characters (as used on checks), and of micro circuits having resonant frequencies peculiar to each material. But such coding is intrinsically contaminating, probably expensive, and is liable to be affected by processing or use.

Accordingly, most effort has been directed at using the coding provided by nature in the form of the physical, chemical or mechanical properties of materials. The "properties" used do not have to be restricted to those tabulated in scientific reference books, such as density and thermal conductivity, but can include such difficult-to-define properties as the noise made by a material when it is hit in a certain way, or the changes brought about when it is stirred vigorously in water.

With so wide a range of possible properties to use for identifying materials, a rich diversity of approaches has been used in attempts to devise practicable central-station reclamation plants having the capability of being substituted for an incinerator or a transfer station in a municipality and of operating at a lower overall cost. Some of the approaches which seem to offer most promise will be briefly described later, after a review of the potential economics of central-station reclamation.

THE ECONOMICS OF CENTRAL-STATION RECLAMATION

Two approaches which might be considered for the economic evaluation of methods of reclamation from refuse in centralized processing plants are as follows. One is to determine both the cost of separating each component and the price which may be expected to be received for this component in various quantities. The wide fluctuations in the price obtained from year to year for waste materials make this line of attack uncertain. The second approach is to consider the cost of alternative methods of waste treatment and, in evaluating reclamation, to disregard any income received for reclaimed materials. The minimum benefit of separating materials is, therefore, that disposal is at least free. This assumption seems sufficiently conservative to predict that if a reclamation plant can prove to be economically advantageous to a community in these circumstances, in practice and at most periods the profitability will be high. The approach of neglecting income is, in the following very approximate analysis, used first, and the possible effects of receiving payment for separated materials is added later.

With income from sale of materials being neglected, the benefit of centralized separation and reclamation has to be limited to the credit for disposing of the community's solid waste. This credit has to be assessed in terms of the alternatives available. The two most-used methods of treatment of solid municipal waste are dumping and incineration. The costs per ton for treatment (not including collection) vary from about \$1.50 to over \$20. The lower figure is for a rural community having within its borders spoiled land of no economic or ecological value until it is filled, and the costs are merely for fencing, supervision and bulldozing. It seems probable, and certainly desirable, that open dumps will be outlawed soon in most communities, and the costs for some form of sanitary landfill, about \$2.50 per ton, will form the lower limit.

At the other end of the scale are those large cities which have used up all available dumping space within their borders, whose neighbors are either outlawing, or charging a high price for, dumping in their territories, and which therefore install incinerators to reduce the mass of material to be dumped to 10 to 20 percent of the initial amount. Incineration costs (as determined in Chapter 7) are from \$5.00 per ton for large (more than 500 tons per day) very well-run plants to over \$20 per ton for small mis-managed plants, and these figures are likely to increase rather than decrease as the costs of improved air-pollution-control equipment increase. At some point in this cost range a separation-reclamation plant should become competitive even with no income being received from sale of the salvaged materials.

Let us examine a hypothetical large-throughput separation-reclamation plant designed for 600 tons per day input on a two-shift, five-day-per-week basis, and actually handling a mean input of 500 tons per day. Suppose that the operating costs (1971 levels) are made up of the following:

Labor	
1 manager	\$ 25,000 per year
1 assistant manager	17,000
1 secretary	8,000
2 supervisors	30,000
2 sorting operators	16,000
2 weigh-scale clerks etc.	18,000
4 baler operators	40,000
2 bulldozer or crane operators	24,000
Total labor	\$178,000
Fringe benefits, pensions, insurance	52,000
Part-time help for sickness, vacations, etc.	30,000
Total	\$260,000

Power costs — assume 150-KW mean load at 15 mills per KWH for approximately 4,000 hours per year	$ 9,000
Heating, insurance, maintenance agreements, taxes, and so forth	14,000
Repairs	30,000
TOTAL ANNUAL OPERATION COSTS	$313,000
Operating costs per ton (500 X 5 X 52 tons)	$2.40

These operating costs do not include the disposal charges for the material remaining after separation of salvageable components. These costs will be examined on a parametric basis.

Let the disposal charges be $D per ton, and let the proportion of input which has to be disposed of as waste be P, (that is, (1–P) tons received are separated and reclaimed). If the operating costs are $L per ton and the capital costs for the plant are expressed as $C per ton, the criterion that it is better to operate the reclamation plant than to go directly to disposal is:

$$\$L + \$C + P \times \$D \ngtr \$D$$

or

$$\frac{L + C}{D} \ngtr (1 - P)$$

This relation, which is also shown in Figure 8.2, gives the break-even condition, which would not of course form an attractive risk for private capital. On the other hand, it is highly unlikely that there would be no income from the sale of any of the reclaimed materials.

Another unaccounted benefit is that it will probably be easier and cheaper to dispose of the remaining material after paper, cans, textiles and cardboard, for instance, have been removed. To give an example of what may be inferred by the above figure, suppose a plant included some mechanization in the sorting operation and was thereby able to reduce the proportion of material ultimately disposed of to 50 percent, with the above operating costs. If the local disposal charges were $10 per ton, and if money for this process could be obtained on the municipal-bond market for 5 percent interest, and for a 15-year plant life with no salvage value, the break-even capital cost of the plant would be $3.5 million.

Since paper alone accounts for between 50 percent and 60 percent of municipal refuse, the figure of 50 percent reclamation should not be difficult to achieve. Certainly this is the case if the dirty paper and garbage are composted, but the incremental capital and operating costs of the composting operation are high, and the probability of finding a market for compost is, unfortunately, low. However, it should be safe to conclude that in areas where the disposal charges are in the region of $10 per ton and where there is the possibility of aggregating a waste flow of at least 500 tons per day (conditions which are found in many large cities) a sorting and reclamation plant should be profitable even if the market for "secondary" materials, including compost, is uncertain.

There are examples to indicate that this proviso, based on assuming that the separated materials will be taken away without payment being made, is too pessimistic. The MetroWaste plant at Houston (described later) was reportedly near to breaking even in 1970 although the

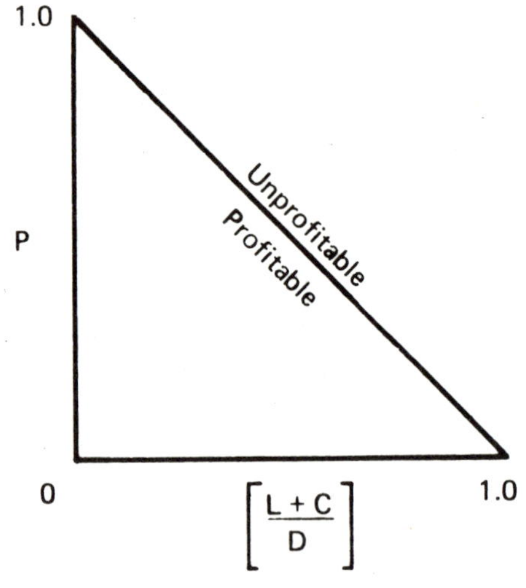

Figure 8.2 Economics of reclamation without income.

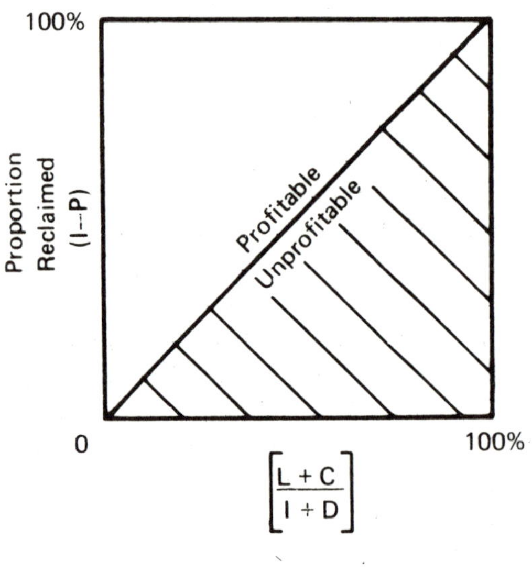

Figure 8.3 Economics of reclamation with income.

amount paid by the city of Houston to MetroWaste was only about \$4.25 per ton. In a higher-disposal-cost area, and with greater automation, profitability should be assured. Sanitary Refuse Collectors of Montreal, as reported by W. J. Johnson [6] made a profit of \$0.57 per ton from salvage by a traditional manual "picking" procedure from a conveyor belt feeding a grinder.

If income from the sale of separated materials is included in the evaluation, the expression for profitability becomes

$$\$L + \$C + P \times \$D - (1 - P) \times \$I \ngtr \$D$$

where the mean income per ton of separated material is \$I. Table 8.1 shows recent (1971) ranges of secondary-material prices. If average prices could be used to estimate the income from paper, cans and bottles a high-capital-cost plant could be justified in areas where the cost of alternatives is high. Compared with an incinerator, a reclamation plant needs no air-pollution-control equipment or a stack, and unless very sophisticated sorting equipment is required the capital cost should be lower than for an incinerator to deal with the same flow.

The importance of reclaiming as high a proportion of materials as possible should be emphasized. If the high cost of installing equipment to handle 600 tons per day of refuse is to be justified, a major portion of this flow must be diverted, and must not simply pass out of the plant to be dealt with by normal methods.

After a review of some of the developments in machine-assisted and fully mechanized sorting, some suggestions for the design of a municipal reclamation plant will be made.

TABLE 8.1

VALUES OF RECLAIMED MATERIALS
IN DOLLARS PER TON
(Chicago)

Ferrous Scrap	Low	High
No. 1 hvy melting, (indus)	\$36.00	\$44.00
No. 1 hvy melting, (dealer)	32.00	43.00
No. 2 hvy melting	29.00	35.00
No. 1 factory bundles	42.00	54.00
No. 1 busheling (Ind.)	34.00	45.00
No. 1 busheling (dealer)	31.00	38.00
No. 1 bundles (dealer)	34.00	44.00
No. 2 bundles (dealer)	20.00	28.00
Machine shop turnings	17.00	20.00
Mixed borings and turnings	18.00	20.00
Shoveling turnings	18.00	22.00
Cast Iron Borings	24.00	26.00
Low Phos & Punchings	45.00	52.00
Heavy forge bars	45.00	52.00
Cut structurals, 2 ft. & under	42.00	49.00
No. 1 ry. scrap, 5 ft.	34.00	46.00
Rails, 2 ft. & under	50.00	57.00
Rails, 18 in. & under	51.00	60.00
Re-rolling rails	52.00	58.00

TABLE 8.1 (continued)
VALUES OF RECLAIMED MATERIALS
IN DOLLARS PER TON
(Chicago)

Ferrous Scrap	Low	High
Steel car axles	50.00	54.00
Car wheels 42 in. & under	43.00	54.00
Angles & splice bars	45.00	54.00
Couples & knuckles	44.00	46.00
No. 1 machinery cast	47.00	53.00
No. 1 cupola cast	40.00	46.00
Clean auto cast	48.00	52.00
Stove plate	32.00	38.00
STAINLESS STEEL		
18-8 bundles, solids	240.00	330.00
18-8 turnings	175.00	250.00
430 bundles & solids	65.00	80.00
430 turnings	25.00	30.00

SCRAP METAL

Dealer's buying prices
(in cents per pound) except where noted

CHICAGO

No. 1 hvy copper & wire	35.00	44.00
No. 2 hvy copper & wire	33.00	40.00
Light Copper	30.00	36.00
No. 1 composition	30.00	36.00
Composition borings & turns	29.00	34.00
Heavy yellow brass	21.00	25.00
New yellow brass clips	28.00	35.00
Yellow brass rod ends	22.00	28.00
Yellow brass rod turnings	20.00	28.00
Auto radiators	21.00	28.00
Heavy soft lead	6.50	7.50
Battery plates, as is	3.50	3.75
Type metal	10.50	11.00
Old zinc and die cast		3.00
New zinc clips	5.00	6.00
Block tin pipe	110.00	120.00
No. 1 pewter	80.00	90.00
Mixed babbit	9.75	10.50
Mixed aluminum clips	10.00	13.00
Aluminum sheet and cast	8.00	11.00
Aluminum borings & turnings	5.00	8.00
Monel scrap	40.00	70.00
New Monel clips	50.00	75.00
Nickel clips	70.00	100.00

TABLE 8.1 (continued)

VALUES OF RECLAIMED MATERIALS
IN DOLLARS PER TON
(Chicago)

Ferrous Scrap	Low	High
PAPERSTOCK in dollars per ton (Chicago)		
No. 1 mixed office waste	$ 1.00	$ 3.00
No. 1 super mixed	5.00	6.00
No. 1 news	10.00	12.00
No. 1 old kraft	25.00	26.50
Dbl. lined kraft corr. cuttings	18.00	20.00
Old corrugated boxes	16.00	18.00
Boxboard clips	2.00	6.00
No. 1 books & mags. grd. wd. free	3.00	7.00
Books & mags mixed	2.00	5.00
White news cuts No. 1	20.00	35.00
White news cuts mixed	30.00	35.00
Ledger stock No. 1 white	25.00	27.50
Ledger stk. col. repacked	18.00	20.00
Manila tab cards No. 1	45.00	50.00
Col tab cards	30.00	35.00
Mixed corr. clips	14.00	16.00
No. 1 hd white env. cuts	50.00	55.00
No. 1 hd white shavings	37.50	40.00
No. 1 soft white	32.50	37.50
New man. env cuts gr wd free	45.00	50.00
New manila env cuts w gr wd	35.00	37.50
Mill wrappers	7.00	9.00

These figures have been extracted from *Waste Trade Journal* for the period October 1970 – March 1971

REVIEW OF SEPARATION APPROACHES

So many attempts are being made to develop methods which will permit central-station reclamation plants to be designed and operated that it will not be possible to mention all. However, what is hoped will be a universal classification of the various approaches will be made first, following which the more promising of the present known developments will be briefly reviewed.

Classification of separation methods

Separation is sometimes one process – for instance, an operator picks a bundle of newspaper from a belt and puts it into a hopper – and sometimes it consists of two processes, coding and switching. An example is in mail sorting, where an operator codes the destination of each parcel, and it is subsequently switched manually or automatically. The distinction between these two possibilities is found in solid-waste separation, and is made in the following classification.

Human-operator separation
Human-operator coding and machine switching
Machine separation
Machine coding and switching

Some of the possible alternative combinations are omitted because of their obvious unattractiveness.

The two classes of machine sorting may be further broken down into sub-categories.

The refuse may be unmodified
Various forms of treatment may be used before, or as part of, the separation, or coding-and-switching process

The range of treatments which may be used is, of course, vast. Those that seem to have significance at present (new approaches are continually being made) are as follows.

Mechanical shredding
Mechanical rolling
Burning in air (incineration – which is a form of separation)
Heating in the absence of air – pyrolysis
Heating to molecular-dissociation temperatures in a plasma (AEC process)
Heating in contact with other gases – for instance, steam and carbon monoxide
Cooling to cryogenic temperatures to employ changed physical properties for separation
Pulping and shredding in water
Reacting with liquid reagents
Fermenting or other bacteriological treatment

The manner in which components of solid waste are identified and separated may also be classified as to whether the process is binary, with successive binary separation processes producing either more components or greater purity of selected components, or multiple-choice, defined as the case where one process produces several separated components. In addition, in some processes each individual item or particle is identified and separated, while in others the material is broken down by shredding or other means to be treated in bulk.

Human-operator Separation

Existing separation plants, or separation stages before incineration or shredding, are generally of the type where human operators stand beside a slowly moving belt carrying the refuse, with each man having responsibility for removing one type of component (clean newspaper, corrugated cardboard and so forth) and for depositing it into a hopper. On the floor below will be, usually, a baler, or the material may be deposited directly into a container or rail car (for instance, in the case of cullet – broken glass – or steel cans).

A rotating-drum or belt magnet is normally used to remove the ferrous material automatically – machine separation – sometimes before and sometimes after other separation processes have taken place (the steel will take with it sheets of paper, cardboard or plastic if it is removed magnetically while mixed with these materials).

Probably the two most prominent – certainly the largest – recently started separation plants relying principally on human-operator (and magnetic) separation are the Lone Star Organics plant of Metropolitan Waste-Conversion Corporation in Houston, Texas and Sanitary Refuse Collectors in Montreal, Quebec. Both were idle at the time of writing. They faced twin difficulties. They relied largely on human operators for the reclaimed materials of value (cans removed by the magnetic separators do not command a high price); and they were located in regions where the cost of alternative disposal was below $5 per ton.

MetroWaste (Lone Star Organics) had introduced a paper-separation machine and a "polishing" (decontaminating) subsequent stage, both apparently of proprietary design, but it seemed that the overall plant economics had not been improved at the time of shut down (for negotiations on a revised price for treatment) to the point where a profit could be made at $4.25 per ton. In addition, much of the investment was tied up with the composting operation, which yielded little income. It is described in Chapter 9.

Sanitary Refuse Collectors closed its separation-grinding plant because local regulations on dumping were suddenly relaxed, changing a medium-cost-disposal area into one where low-cost dumping could be employed by competitors.

These details are given because of the interest in the potential viability of traditional types of reclamation plants and of plants which promise greater operating economies through the use of mechanization to some extent.

Human-operator Coding and Machine Switching

In human-operator separation, not only is the speed of separation slow, but a separate operator is needed for each component. This is very poor utilization of human brain, which has a very high processing rate. A typist or a pianist, to take two relevant examples, can handle 30 "bits" of information per second. This rate would give, for example, the ability to sort three items per second into ten categories. This is beyond the rate at which we can presently handle solid waste in such a way that we could present each individual item for coding.

At the time of writing, a pilot plant is being constructed at M.I.T. to use human-operator or machine coding. Its design owes much to the work on the automation of post-office sorting [7, 8] and airport baggage handling [9].

The refuse is to be fed on to a moving belt by conventional arrangements, and an "over-band" or rotating-drum magnet will remove ferrous material for baling (Figure 8.4). A vibrating sieve would then allow all pieces of smaller than, say, a quart soda bottle to drop away for treatment in an air classifier (described later). The remaining large items will be fed to a fast-moving carousel (a belt could be used equally well) in such a way that each item is separated from the next, in this case by partitions. This is the most uncertain part of the process. It is a normal step in materials handling, though the highly nonhomogenous nature of mixed refuse will undoubtedly require special treatment. The carousel (or belt) will pass in view of an operator at a coding station, who will have before him a console having buttons. If there are less than ten categories, as in the present design, each button is associated with one category. More categories can be given by using combinations of buttons. By pressing appropriate buttons as each piece of refuse appears the operator will be able to code a magnetic "strip," one for each compartment. If the code indicates "glass" – or if the economics indicate that it is worth sorting glass by color, "green glass" – the coding would later actuate an appropriate diverter to sweep the glass into a hopper. Most likely the glass would be ground or it might be sorted by color mechanically before being ground, or whole bottles might be separated for return to the bottlers if the incremental value justified the operation.

In any event, the productivity of the human sorter is increased 5 to 10 times by this mechanical assistance. Furthermore, it will be easier to hire reliable operators for refuse sorting when a white-collar environment can be offered, and it will be justifiable to pay a higher reward in addition to an incentive bonus based on reclaimed value of materials. It seems probable that this type of plant will form an intermediate step before full machine coding and sorting is practicable.

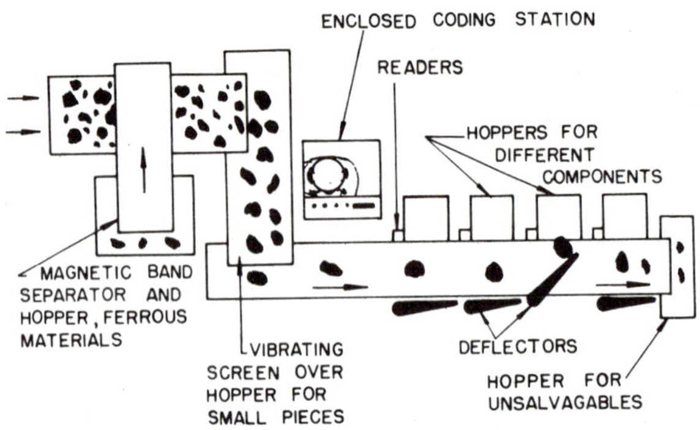

Figure 8.4 Proposed plant with human coding and mechanical switching.

MACHINE SEPARATION

Sorting by Pulping

A method of separation which reached the pilot-plant stage in about 1969 is the so-called "Hydropulping" process of the Black-Clawson Company, [10], manufacturers of paper-making machinery. A full-scale plant is due to come into operation in Franklin, Ohio in April, 1971. By passing all municipal refuse into a little-modified pulping tank, in which pumps and impellers induce a strong vortex and associated mixing and tearing, the paper products break down and are recovered by filtering the resulting pulp through a succession of perforated-screen and slotted classifiers and cyclone separators, (Figure 8.5). A 5/8-inch-dia-hole screen is used first to take out the glass, cans and so forth and subsequently a screen with 1/8-inch-diameter holes and another with 0.014-inch slits are used to select fibers and to reject flakes, rubber-like balls (from, for example, pressure-sensitive adhesive backings) and pieces of plastic film. Subsequently a close screen (150-mesh wire) is used to retain the desired fibers and to reject the fines. The fiber is dewatered and thickened and baled for use in a paper mill.

The ferrous cans are recovered magnetically and the glass from bottles broken in the pulper is screened out.

The Black-Clawson process is thus a complete binary branched system. (There are additional binary sorters on the metal and glass branches designed to yield products with as little contamination as possible). Over 50 percent of the incoming feed is sorted into reclaimable materials. Including income from these products, economic predictions made by Black-Clawson are for refuse-treatment costs of 70 cents per ton for a 500-tons-per-day plant and a

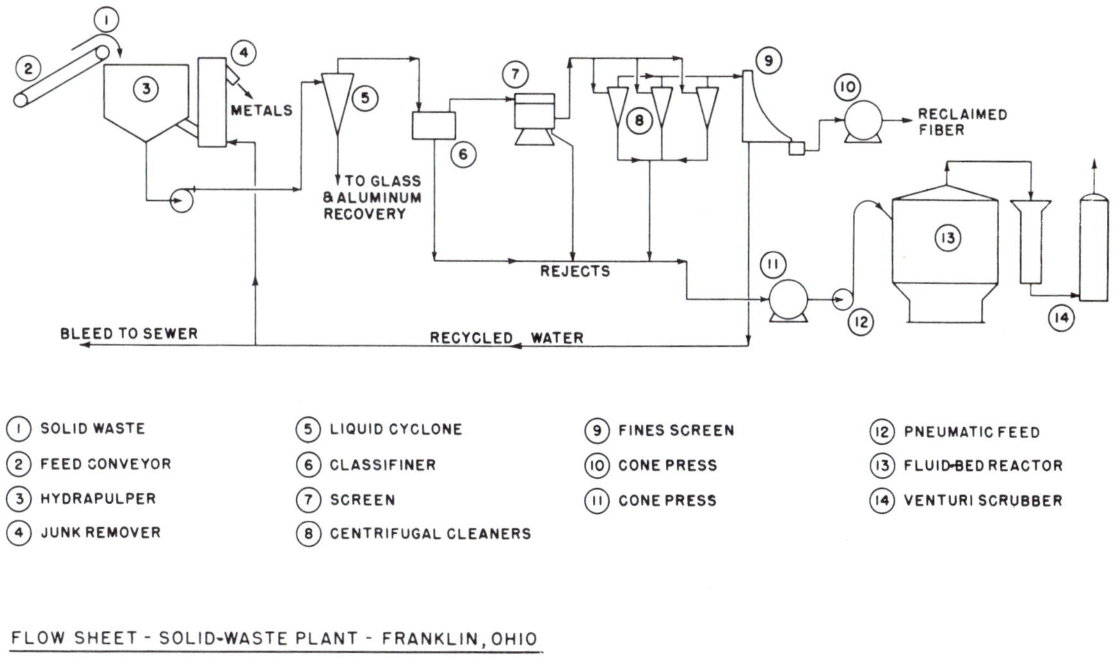

Figure 8.5 Black-Clawson hydropulper process.

net income of 90 cents per ton if the volume is 1,000 tons per day. If these projections are borne out in practice the whole solid-waste picture would be transformed, and steady markets for secondary pulp, glass cullet and cans would need to be assured.

Sorting by Ore-classification Methods

The Bureau of Mines is developing at least two approaches to sorting and reclamation, one operating on incinerator residue and the other having 'raw' refuse as feed. In both cases the philosophy is to regard refuse as a low-grade ore, and to apply to it only slightly modified ore-classification processes.

Few details have been released regarding the work on the raw refuse except that air classification will be employed, but the incinerator-residue plant [11] has been in the pilot-plant stage since about 1969. The flowsheet, Figure 8.6, shows a layout of approximately 14 binary separations consisting of continuous screening, shredding, magnetic separation and gravity concentration. A diagram of the plant is shown in Figure 8.7. The addition of further binary "polishing" stages, such as the Sortex separation of cullet into colors, (described below) is possible. The capital cost of a plant to process the incinerator residue from 1,000 tons per day of raw refuse was estimated by Stanczyk and Ruppert [11] as about $1.75 million and the operating costs as about $2.80 per ton. Neither the costs of disposal of the fraction for which use cannot be found nor credit for the useful components were included in the operating-cost estimate.

Another approach to resource recovery from incinerator residue is being investigated at the Illinois Institute of Technology. Blum, [12] shows that certain metallic and glassy components separate out when the residue from a high-temperature (slagging) incinerator is cooled in a certain manner (Figure 8.8).

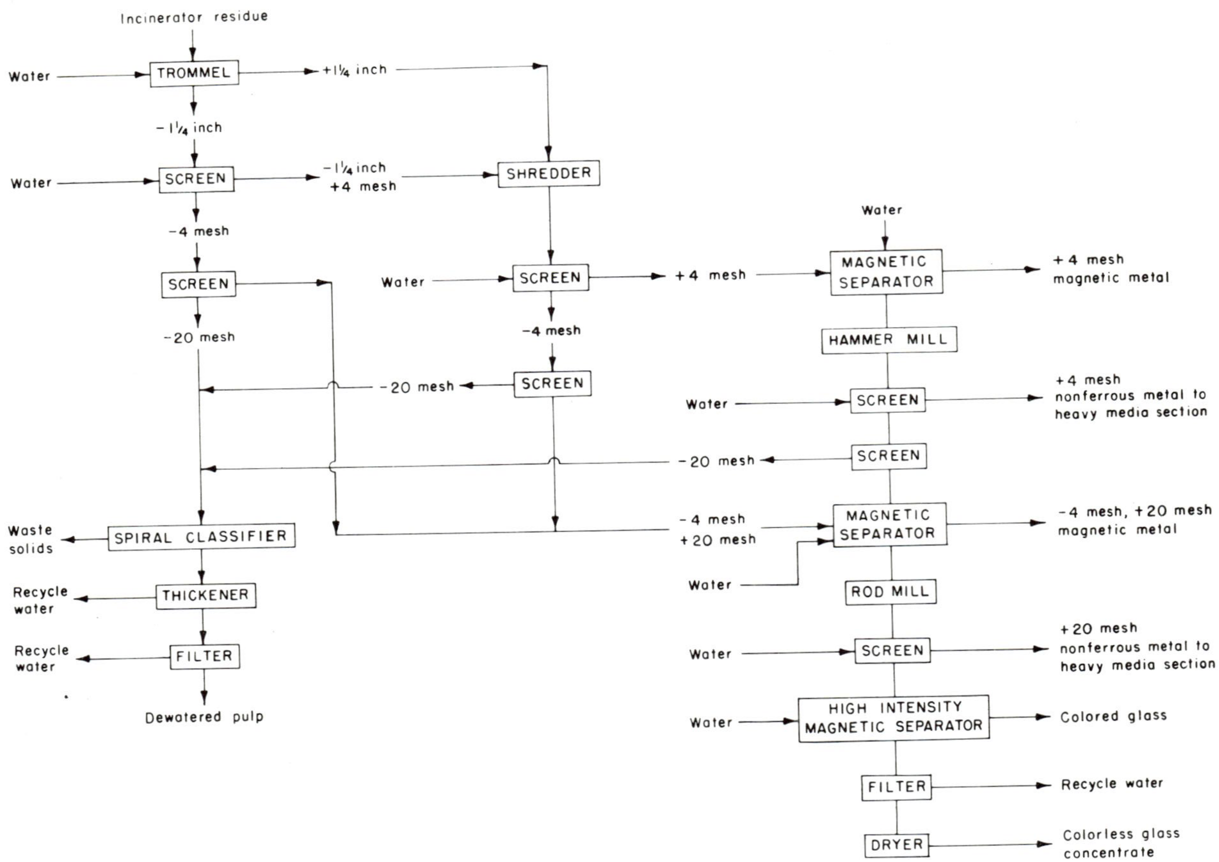

Figure 8.6a Physical beneficiation flowsheet for concentrating valuable materials.

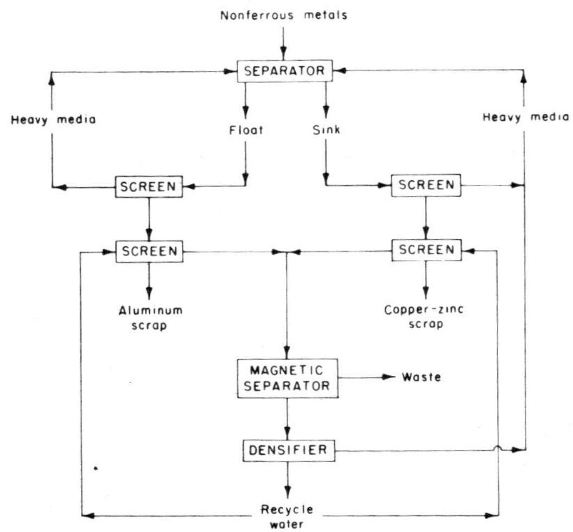

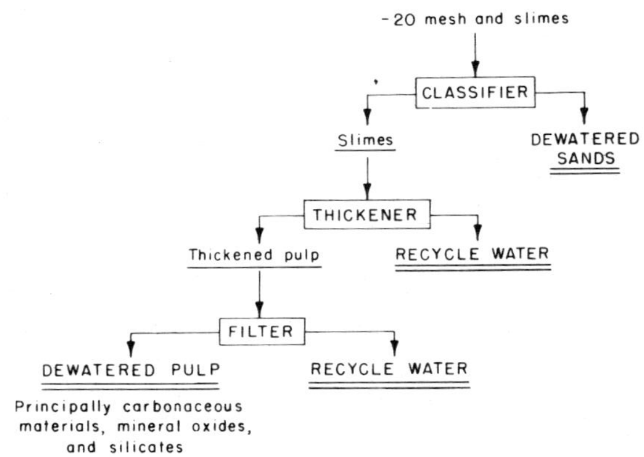

Figure 8.6b and c. Bureau of Mines process for reclamation from incinerator residue.
courtesy U.S. Dept. of Agriculture

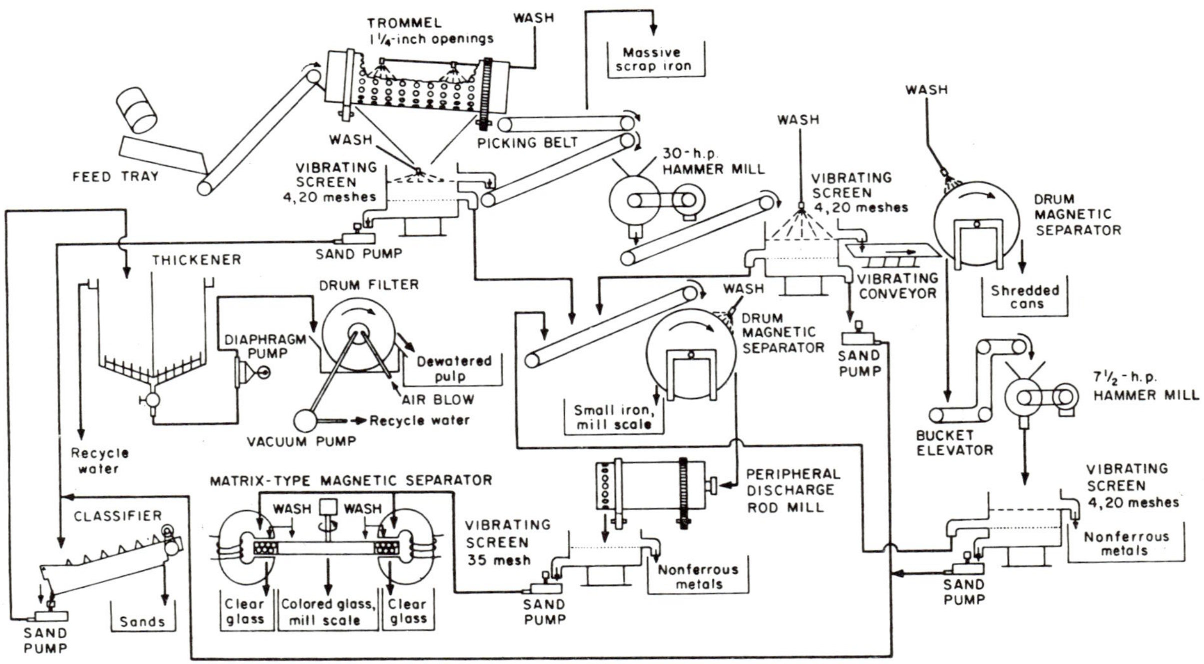

Figure 8.7 Representation of Bureau of Mines process.

courtesy U.S. Dept. of the Interior

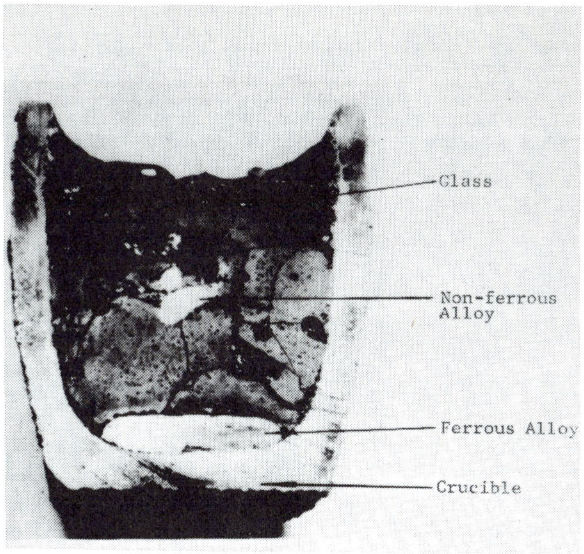

Figure 8.8 Separation of components from incinerator slag.

courtesy IITRI [2]

Color Coding and Air Switching

A machine has been developed by the Sortex Company of North America to sort translucent ore by color, (Figure 8.9). It is being used at the Bureau of Mines to sort glass fragments into flint glass and colored glass, and by appropriate use of filters a series of such machines could sort glass into the four principal colors of bottle glass. The glass pieces of roughly uniform size (obtained by initial screening) are fed from a hopper to a vibrating table and thence to a grooved belt moving at 5 to 10 fps. As the pieces fall from the end of the belt each one is sensed by three photocells appropriately filtered, and the smaller fraction, colored or uncolored, is diverted by rapid-acting air blasts to a separate hopper.

Conductivity Coding and Air Switching

Instruments which sense the change of electric-field strength when a conductor is moved through the field can, as with mine detectors, discriminate between metals and nonmetals. So far as is known, such instruments have not been used in refuse-reclamation plants for two reasons. Firstly they code only, and sorting and switching mechanisms would have to be designed just for this one operation. Secondly, mixed non-ferrous metals are of little value; means would have to be developed to separate copper from brass from aluminum and so on. It may be inferred from descriptions below of other processes that means might in fact now be available.

Automatic sorting by direct measurement of conductivity has been developed to an advanced stage in the ore-machinery industry. Figure 8.10 shows diagrammatically a machine produced by Ore Sorters (Canada) Limited, a subsidiary of Rio Tinto Zinc Corporation, in which the feed is distributed on to, and falls down, a 48-inches-wide angled slide. One or more

Figure 8.9a Sortex color-sorting machine.
courtesy Sortex Company of North America, Inc.

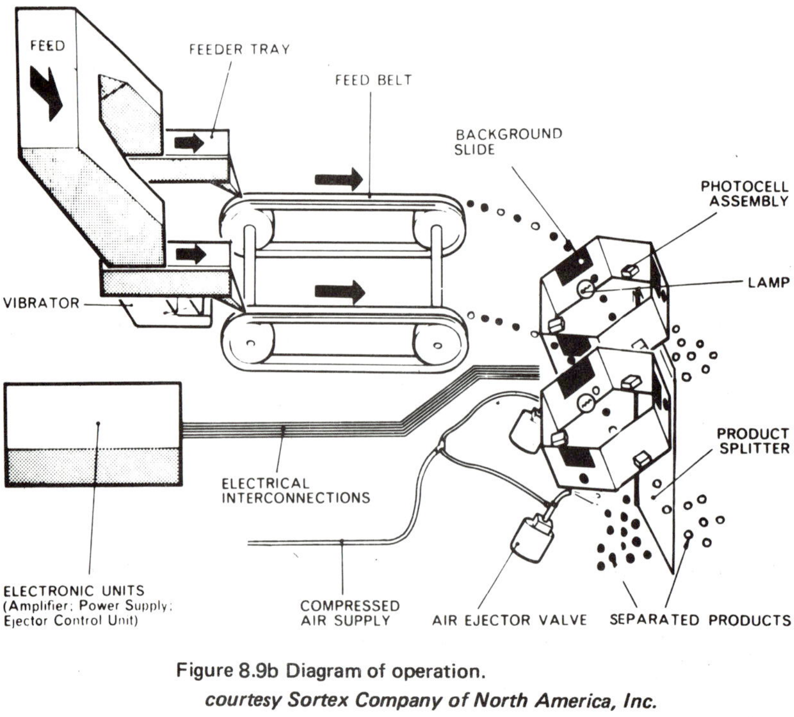

Figure 8.9b Diagram of operation.
courtesy Sortex Company of North America, Inc.

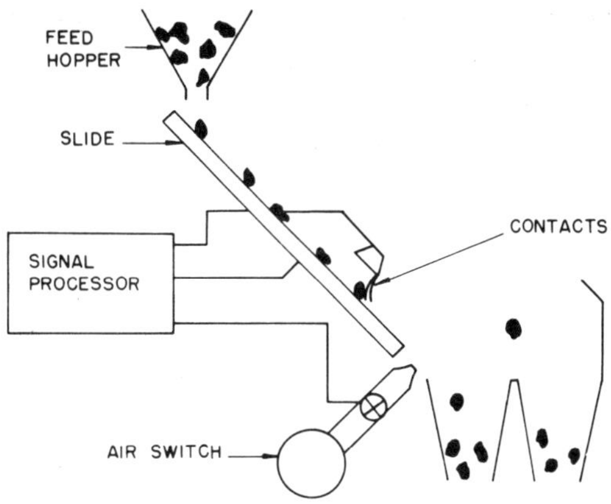

Figure 8.10 Ore sorting by conductivity.

courtesy Ore Sorters Limited

of a series of electrodes carrying a steady potential with respect to the grounded slide make contact with each particle, and the current through the particle enables it to be coded as "high" or "low" conductivity. As the pieces fall off the end of the slide, air blasts divert either the high-or low-conductivity pieces, whichever is in the lower proportion.

Photometric and Radiometric Coding and Air Switching

Ore Sorters Ltd. has also developed machines which sort on the basis of surface reflectance of a scanning laser spot, and of radioactive properties. The photometric unit may be of value in application to mixed refuse, and it is hoped that trials will be made. A random-feed stream is distributed on a horizontally moving conveyor belt and is scanned by a laser spot reflected from a rotating mirror, Figure 8.11. A photomultiplier measures the characteristics of the reflected light, and an analyzer determines whether or not appropriate air jets should be actuated to divert each piece.

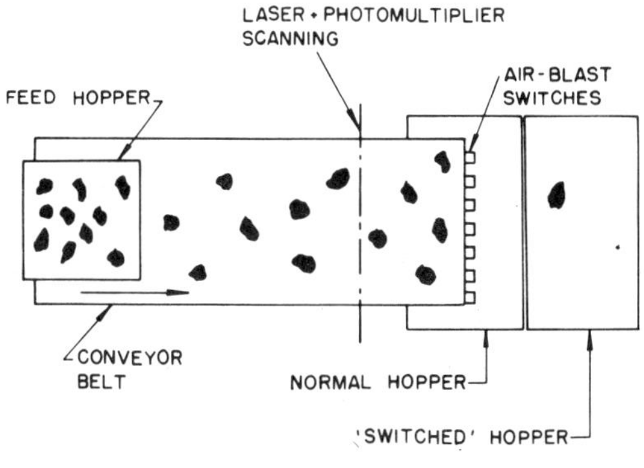

Figure 8.11 Ore-Sorters' photometric sorter.

courtesy Ore Sorters Limited

X-ray-attenuation Coding and Mechanical Switching

Another industry which has developed coding and sorting machinery to a high degree is the agriculture and food industry. An example of the use of x-rays for coding is shown by a machine conceived by a Scottish farmer for separating potatoes from stones and dirt [13]. The principle is similar to that of the two belt sorters just described except that switching is by movable fingers rather than by air blast (Figure 8.12).

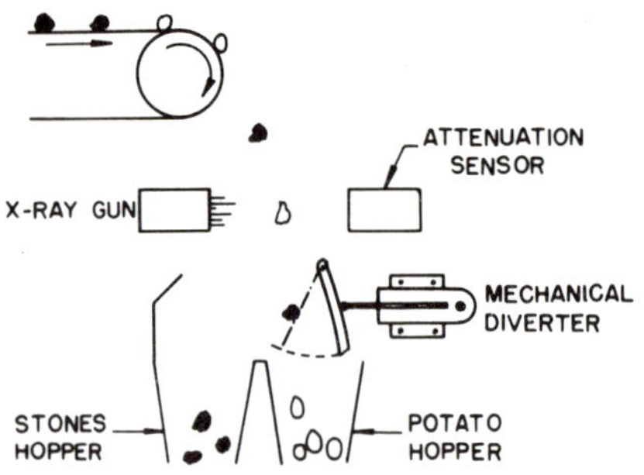

Figure 8.12 Sorting by X-Ray attenuation.

Sorting by "Bounce"

A well-established method of sorting hard from soft materials is to allow a stream to fall on to a moving, and perhaps vibrating, belt.

Soft materials are carried away by the belt. Hard materials bounce off the belt into a hopper. This type of arrangement is used by MetroWaste to separate the larger ground-glass particles from compost.

Cryogenic Sorting

A news release is all the present information available about a Belgian process – the George process – which relies on the change of physical properties of many materials when cooled to a low temperature.[1] Rubber, for instance, becomes hard and brittle and can reportedly be easily handled by a hammer mill (automobile tires stall all but the largest hammer mill at normal temperatures), after which it is claimed that separation by density or other physical properties is somewhat easier than at room temperature.

Separation by Density and Drag

S.R.I. Zig-zag Air Classifier. – A machine developed for the food industry, principally for the sorting of food beans, has been applied by Stanford Research Institute to the separation of mixed municipal refuse, [14]. The so-called air classifier consists of a zig-zag duct arranged with its principal axis vertical (Figure 8.13). Air is drawn off from the top and the mixed material is fed into the duct at an intermediate point. Pieces with a combination of low density and high drag are entrained into the rising air stream, while the heavier or lower-drag pieces descend. It is claimed by S.R.I. that components having densities only a few percent apart can

[1] A later first-hand report was to the effect that a plant using the George process was in full commercial operation on scrap automobiles in 1971.

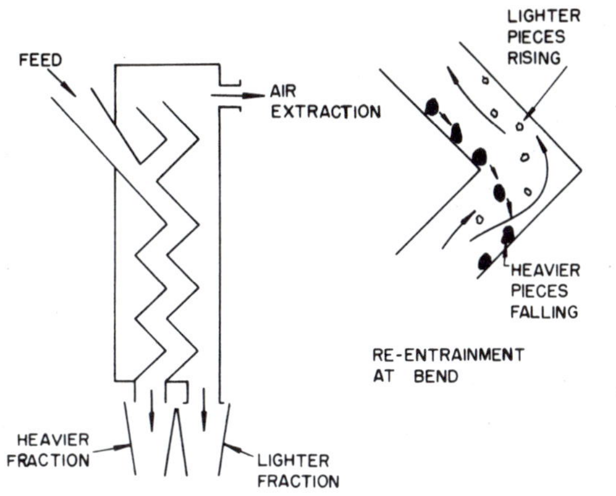

Figure 8.13 Stanford Research Institute zig-zag classifier [14].

be successfully and economically sorted. If this claim is substantiated, the machine could obviously have a very beneficial impact on the reclamation picture. This type of machine requires the input to be pulverized, dried and screened so that any one classifier is fed with pieces varying in size by only a small degree. Tests have been made using polyethylene spheres, but it is in the treatment of nonspherical particles – especially pieces of sheet material – that the usefulness of air classification will become known.

Fluidized-bed Separator. – A fluidized-bed separators has been adapted for pulverized refuse at the Warren Spring (Ministry of Technology) Laboratory in Britain [15]. Air is supplied to several sections of a porous-bronze table, which is vibrated, Figure 8.14. A size-selected fluidizing medium, in this case ferrosilicon or magnetic or iron powder, is kept in suspension and recirculated, the feed particles being larger than those of the medium so that filtering can be used. The denser particles, which sink to the bottom of the bed, are moved by the vibrating table up the bed, while the lighter feed particles flow downwards along the top of the bed. This unit has been installed in a pilot plant. Operating costs of 25-50 cents per ton have been forecast, including depreciation.

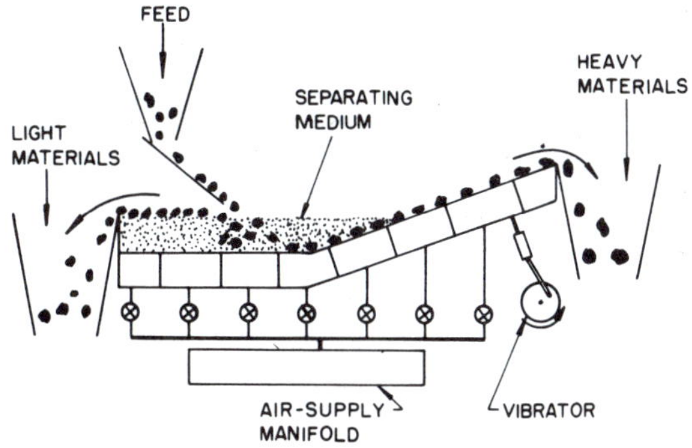

Figure 8.14 Warren spring fluidized-bed classifier.

Vortex Multi-output Classifier. – This machine, which has its counterparts in the centrifugal separators of chemistry and medicine, sorts in one operation by sensing a combination of density and drag, as for the zig-zag classifier. However, instead of the input stream being separated into two output streams of "high" and "low" density the material is distributed in a continuous manner over a range of density-drag combinations, Figure 8.15. It is being developed at M.I.T., [16].

The principle of operation is this. Particles are allowed to reach an equilibrium position in a radially inward-flowing vortex. The radially inward component of the flow produces a drag force on the particle which depends on the shape and size of the particle and on the square (approximately) of the inward flow velocity. This velocity increases as the radius is reduced, the variation being dependent on the shape of the vortex chamber. The drag gives a centripetal acceleration to the particle which in equilibrium conditions will be moving in a circular path with no relative velocity to the tangential component of the flow. The classifier is intended for use in batch feed with air or water as the fluid medium. As soon as a batch of particles has taken up equilibrium radii, scoops will lower across the stream to capture particles in groups according to their radii. The scoops are removed and a new batch of particles is added.

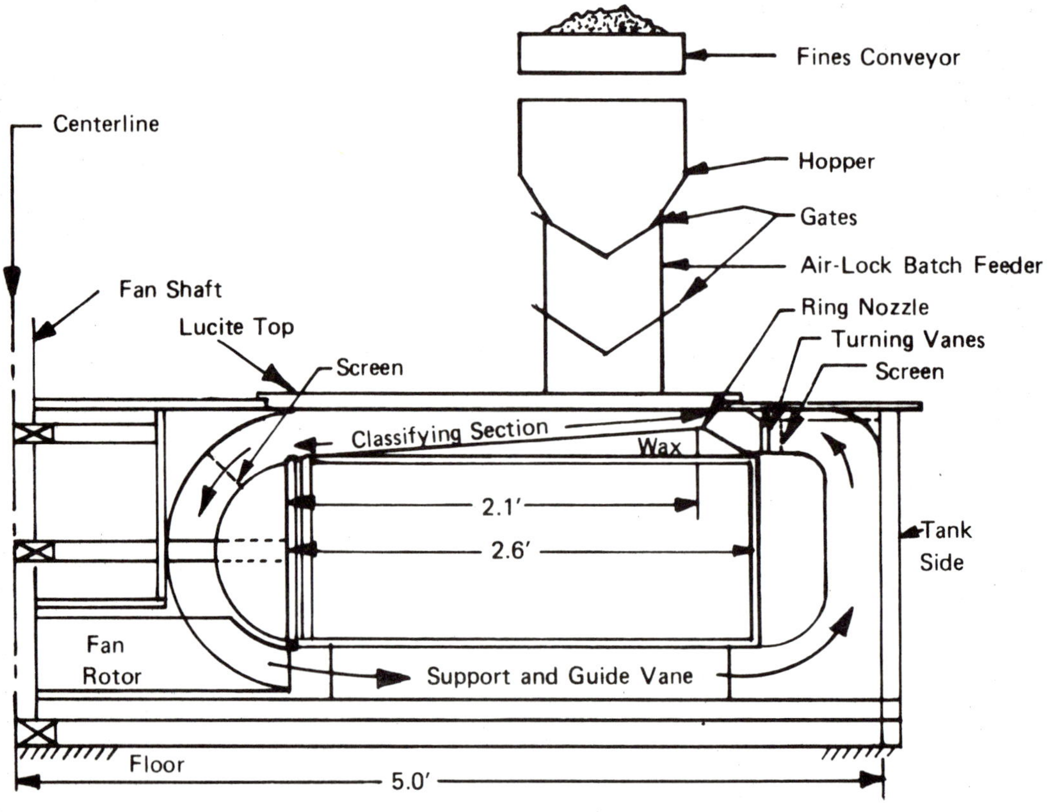

Figure 8.15 Vortex multiple-output classifier.

Machine Coding and Switching in Series

Some examples were given in the previous section of machine coding and switching for binary-choice separation. In this section developments which promise separation into many categories from one coding station (possibly incorporating several sensors) will be reviewed. All activity in this category is, so far as is known, being carried out at M.I.T.

In multiple-choice series systems, the inputs from one or more sensors are analyzed by a central logic unit and a sorting decision made upon the basis of the total input, Figure 8.16. The number of sensors involved is a function of the number of sorted categories desired, the limits to contamination desired, and the amount and accuracy of information available from each sensor.

Many different sensors have been under investigation for possible use in the series system. One concept is that of the "total spectrum" or signature approach.

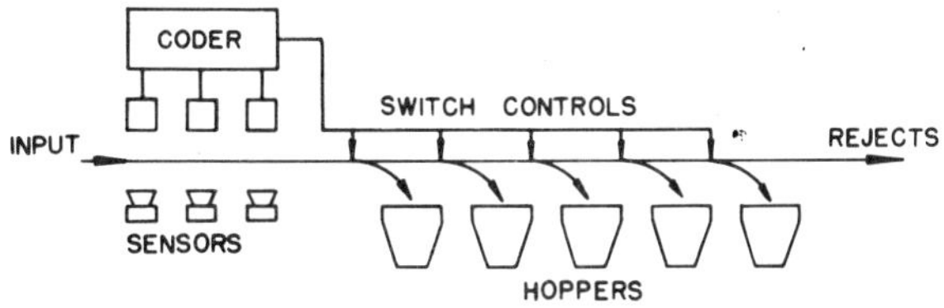

Figure 8.16 Multiple-choice series system.

Signature Systems

Whenever the output of a sensor is a continuous function analogous to some property or combination of properties of a material, the output is referred to as a "signature." Such is the output of an accelerometer attached to some form of impact tool (e.g. a steel ball) which is driven (or allowed to fall) on each item of refuse, Figure 8.17. The resulting impact signature is a time-dependent voltage output from the accelerometer representing the deceleration forces generated when the impact tool deforms the sample. It is clear from visual examination of Figure 8.18 that quite different signatures result from impacting different materials.

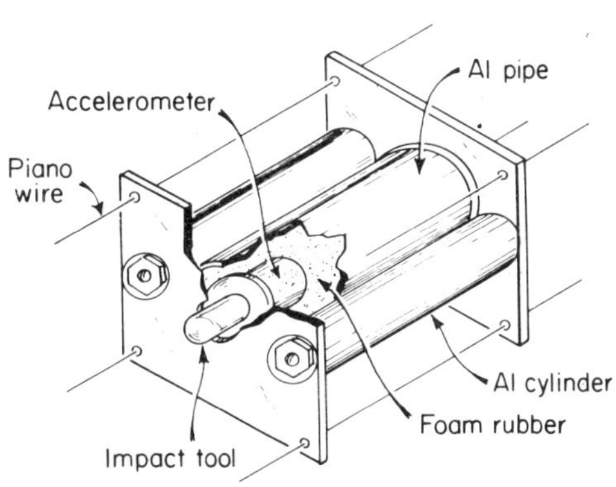

Figure 8.17 M.I.T. impact sensor.

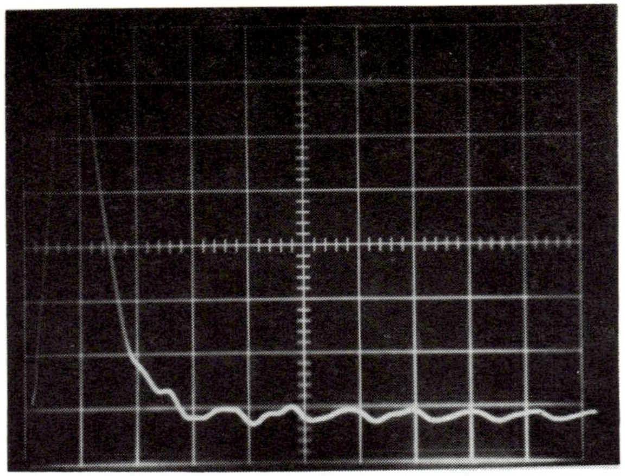

a. STEEL

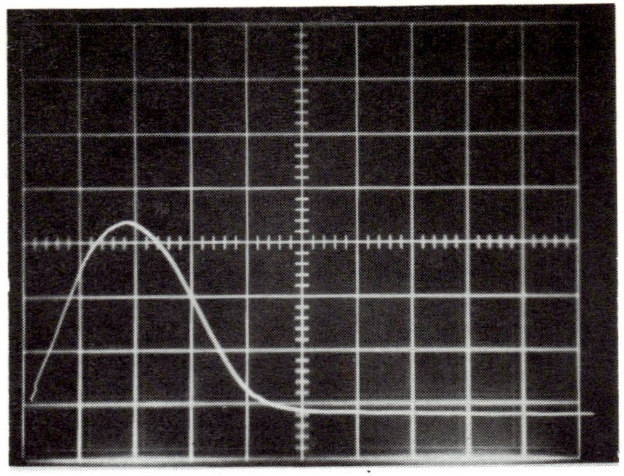

b. GLASS

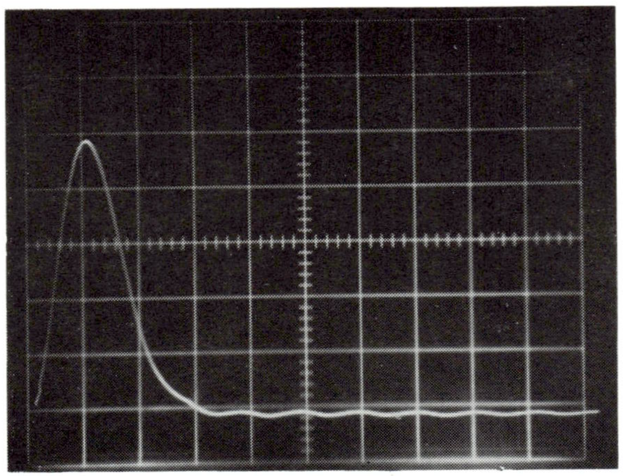

c. ALUMINUM

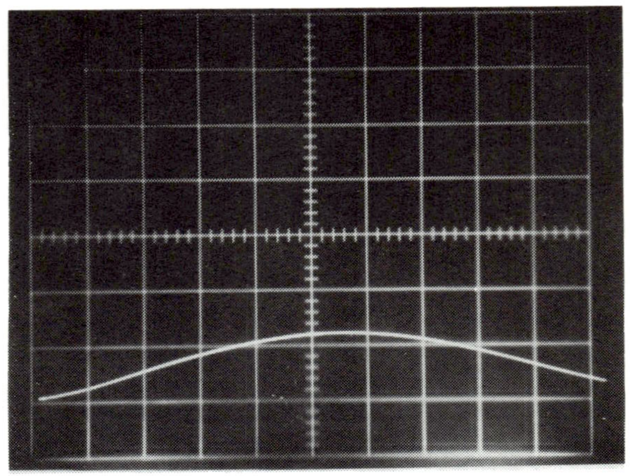

d. WOOD

Figure 8.18 Examples of impact signatures.

It has been found possible to extract a great deal of information from the voltage waveform, so that although the signature of an aluminum can, for instance, will be quite different from that of an aluminum bar, they will both be classifiable by the same coder [17].

Infrared Coding

A sensor which is complementary to the impact sensor is one which investigates the infrared reflectance spectrum of each item of refuse. Of particular importance is the strong indication of whether film material is cellulosic or not from the absorption of a narrow range of wavelengths by the O-H bond in cellulose [17]. This reading gives a strong indication of the material being paper or plastic.

UTILIZATION OF RECLAIMED MATERIALS

The ways in which reclaimed materials can be reused are numerous and are being continually added to. All that will be attempted here is to give some of the principal uses of the main constituents of municipal refuse, to indicate some problem areas, and to refer to sources of further information.

Paper, Corrugated Cardboard

In general, paper and cardboard can be repulped for the manufacture of "new" paper or board. There appears to be no limit under present conditions to the number of times paper can be recycled, because the proportion of paper which has been already recycled and which comes back a second time is small. If there were to develop much more widespread recycling, the proportion of fibers rejected by the pulping mills because they were too small would rise, but there would be no other presently obvious problem. The reject fibers could find other uses, some of which are mentioned below.

The quantity of paper recycled increased from 8,800,000 tons in 1956 to 11,500,000 tons in 1969, about 20 percent of total consumption.

The principal use (70 percent) of paper and cardboard is in the manufacture of combination paper-board, corrugated board and similar materials where color is not of primary concern (Figure 8.19). De-inking processes have now reached the stage, however, where high-quality white paper can be produced from printed newspaper [18]. These processes presently are causing problems in stream pollution in certain areas, and their future growth is dependent on solving these problems as well as accomplishing a reduction in costs in relation to paper produced from virgin fibers. Paper produced from secondary fibers can be more stable dimensionally and have greater opacity than when made from virgin fiber: it is by no means a low-quality, "substitute" product.

The uses of paper and board are governed principally by contaminants, most of which present severe problems of sorting and removal. The sorting is carried out in general at the dealer level, though mills must make spot checks because the costs involved in, for instance, pulping a batch of paper containing some foam-plastic cups or a few sheets of paper carrying pressure-sensitive adhesive can be very large. New equipment is continually being developed to cope with new glues used for magazines and books, or carbon paper, or plastic coverings found on some advertising material and milk cartons, for example. Asphalt and wax give particular difficulties. Magazine paper can contain 30 percent by weight of kaolin. Paper itself is produced by several processes, of which the groundwood process is the least expensive, and groundwood must not be admitted in the production of high-quality papers from paper stock.

Figure 8.19 Seven-ply-paperboard machine – "Ultraformer."
courtesy The Beaver Wood Fibre Co., Limited

The problems caused by these contaminants are such that mills and dealers would prefer to take paper which is intrinsically segregated at source – newspaper overruns, computer tab cards and print paper, envelope trimmings – and the mixed grades of paper often find uses in the manufacture of roofing, drainage-pipe and board materials where plastic, rubber, wax and other contaminants are relatively unimportant.

Mixed paper and board can also be used as a source of cellulose in a number of processes where contaminants may be less troublesome. The production of industrial ethanol from waste paper has received renewed attention following the work of Porteous [19, 20].

The Bureau of Mines is investigating pyrolysis and hydrogenation of refuse [21, 22] with the useful products coming principally from cellulose. One ton of mixed refuse is reported by Corey [21] as producing up to 18,000 cubic feet of gas having a Btu content of 6 million together with some tar and oil, from pyrolysis, while if the same quantity of refuse is treated with carbon monoxide and steam about 2.4 barrels of largely paraffinic oil is produced.

Experiments are also being carried out in several places into the conversion of the cellulose into yeast [23] or other food, primarily for cattle, by bacterial action [24].

Wood

Demolition lumber in municipal refuse is presently buried or burned in most cases. (In New York City and Boston, lumber is loaded aboard barges which are towed to the outer harbor and their contents burned when the wind is off shore.) Problems in reuse arise because of the large variation in size of waste lumber – beams a foot square and sawdust may be in the same batch – because of the contaminants, and because of the wide variety of types and conditions of timber involved. Shredding is necessary in almost all processes presently being considered for wood reclamation, and hammer mills or grinders of very large power levels, of the order of 1,000 hp upwards, are necessary to cope with large pieces of wood. Lower-power chippers, which employ sharp cutting edges, can be used only when it can be guaranteed that the wood contains no metals or ceramics.

Once waste lumber is ground or chipped it becomes easy to burn in an incinerator. Useful products may be produced from pyrolysis, though the process and the economics are as yet uncertain (see Chapter 7). Hydrogenation (mentioned above in regard to paper) is also possible. An excellent review by Harkin of uses for sawdust, shavings and waste chips [25] discusses a large variety of possibilities and of actual present practices, of which chipboard and other types of building board represent the greatest potential. This review was concerned mainly with the waste wood from sawmills and other wood-processing operations, and since these waste-wood products are more homogeneous and less contaminated than are municipal and demolition wood wastes, it is to be expected that a wider variety of opportunities exist than for mixed lumber.

Glass

Glass presents a comparatively simple picture as regards recycling, since its extreme stability prevents it being economically modified except by being mechanically ground or thermally melted. In general, the recovery of returnable or otherwise reusable bottles from mixed refuse, once widely practiced, can no longer be classed as worthwhile. The development of alternative methods of returning bottles to use, such as a bottle-reclamation machine proposed in the original M.I.T. summer study on solid wastes [26], would increase the return flow of whole bottles. At the present time, most glass recycling is in the form of cullet (broken glass).

Glass, particularly bottle glass, can be remelted and reused indefinitely, and bottle manufacturers are able to work with 100-percent cullet as feed [26]. There are contamination problems: the bottles should be of uniform color and should contain no aluminum caps or foil labels. But food and other organic contaminants are not of concern as they burn off in the furnace.

In its ground form, glass has been used for many years in the manufacture of, for instance, roofing materials [26] and its incorporation into asphalt on an experimental basis has created widespread public interest [27]. The Bureau of Mines is producing experimental structural bricks from the inorganic materials in municipal refuse, principally glass [28]. Abrahams summarizes the present position regarding the utilization of waste glass in the same symposium [29].

Credits for reuse of glass must come mainly from the savings in the cost of disposal rather than in the conservation of raw materials, which are low in cost and readily available in virtually all areas.

Ferrous and Nonferrous Metals

Like glass, metals can be repeatedly remelted for reuse virtually indefinitely. However, whereas there are only comparatively minor variations in composition among bottle, window and chemical glasses and a small amount of contamination is tolerable, the presence in a melt of a primary metal of almost any other metal can lead to serious problems. Most metals are produced in batches, rather than in continuous processes, and if the feed to a 300-ton basic oxygen furnace includes copper-bearing, tin-bearing and nickel-bearing scrap which raises the copper, tin and nickel above minimum levels, (typically 0.08 percent copper, 0.02 percent tin and 0.04 percent nickel), the whole melt, worth tens of thousands of dollars, will drop drastically in value.

Ferrous Scrap

Accordingly, metal producers are not keen to accept scrap material unless its degree of contamination can be guaranteed. Steel mills prefer scrap consisting of massive castings to bales of crushed cans, an uncertain percentage of which will be tin coated, soldered, have aluminum ends or sides, and may contain residues of all manner of materials, most of which will be harmless but some of which may be extremely damaging.

For this reason, the principal use for cans is in the catalytic refining of copper, a use for which there is insufficient capacity to absorb a much greater flow of scrap material than at present.

The processes for recycling ferrous metals have undergone, and seem likely to continue to undergo, major changes. When most steel was produced by open-hearth furnaces, up to 45 percent of scrap could be used in the charge. A considerable amount of steel scrap is produced in the whole process of steel manufacture, but the open-hearth furnace was still a net consumer of scrap. The major switch to the basic oxygen furnace over the last two decades has led to steel plants which produce more scrap than they consume, since a BOF can take a maximum of cold scrap of only about 33 percent of the charge [30].

The next development, which has been occurring at an increasing rate over the latter half of the 1960's, has been the introduction of comparatively small electric-arc furnaces, which can operate on 100-percent scrap. A "minimill" needs only about half the capital investment per ton of capacity as a big mill [31]. These small mills are, however, at the mercy of fluctuating steel-scrap prices.

In response to this financial and management problem, a "direct-reduction" process has been developed to reduce iron ore to iron-rich pellets that can be fed directly into electric furnaces. A question which is of importance to the future of steel recycling is whether or not the direct reduction of iron ore in electric furnaces will become more attractive than the use scrap as feed because of either lower delivered prices or more consistent quality or the elimination of uncertainties regarding supplies. Lukens Steel, which makes plate principally out of scrap, found its 1970 profits cut sharply because of scrap prices remaining at "abnormally high levels" [32]. The company realized a 20-percent increase in output and an overall reduction in costs based on an electric-furnace charge of 20-percent pellets and 80-percent scrap [33].

Nonferrous Scrap

As can be seen from Table 8.1, nonferrous metals fetch far higher prices than does even high-grade scrap steel. This means that greater effort can be devoted to sorting and separating. Scrap dealers use accurate methods of assay to ensure that maximum prices are given for high-purity scrap. The low melting point of many nonferrous metals is sometimes used with advantage for separation: for instance, in the Coreco metal separator [34] which can be used to recover first zinc and then aluminum alloys from a feed of automotive fuel pumps and carburetors.

Nonferrous scrap is used directly as feed to remelting furnaces. In the face of predicted world shortages within a comparatively few years of many nonferrous metals , one would reasonably expect a strong future for reclamation. However, at a time when it is said that there is not enough copper presently available to supply the peoples of Africa with a telephone system equivalent to that of the U.S., it is strange to report that there is a world "overproduction" of copper, with the price falling 40 percent, to 47 cents per pound, in a year [35]. This development obviously affects recycling quantities and profitability.

Aluminum is currently receiving much public attention because of can-reclamation drives supported or actually organized by aluminum producers [36]. The Aluminum Association reported that at least 115-million all-aluminum cans were returned to aluminum-producer can-reclamation centers for recycling during 1970 [37]. There is some skepticism about the economic viability of "can-by-can" reclamation, which has reached the point where legislation has been drafted to prohibit reclamation purely for public-relations purposes [38]. However, the obvious fact that all-aluminum cans present far fewer problems for recycling than the various types of steel cans will help to promote legislation favoring aluminum.

Ash and Slag

The reuse of ash and slag is at the boundary of the principal subject of this book, which has concentrated on municipal wastes. Formerly, municipal wastes were principally ashes; now they are principally paper. However, the municipal incinerator should produce ashes and slag, and some of the extensive studies which have been made into the use of power-plant ash for ready-mix concrete, concrete blocks and highway construction are relevant [39, 40, 41 and 42]. In some places, slag piles, which seemed destined to be permanent eyesores, have been "mined." The Maryland Slag Company is processing mounds of black-furnace slag, separating the iron-rich pieces magnetically for feeding back to the blast furnaces, and turning the remainder into an aggregate [43]. Perhaps this is a preview of the future of our municipal dumps.

Tires

Reclamation of rubber from used motor tires ran into difficulties a decade or so ago as the several varieties of synthetic rubbers and cord materials began being used on a widespread basis. Prices quoted by Lipsett [44] for 1962 were $257.60 per long ton for first-quality tire reclaim, and 11.5 cents per pound for reclaimed rubber. Current (1971) prices for mixed tires delivered at Akron, Ohio are ten to 13 cents per ton. Barnett [45] discusses in realistic terms the problems of tire reclamation.

Much interest has been attracted by processes which convert scrap rubber tires to products other than rubber. Several companies are experimenting with pyrolysis plants [46] which break down the rubber into combustible and other gases, oils, tars and waxes (Chapter 7). Ground rubber from old tires has also been used as a filler in paving mixes and as a source from which to produce carbon black to use in new-tire production [47]. None of these uses has yet become established as a major method of used-tire reclamation.

Plastics

There is much in common between plastics and tires, because they both come in so many formulations that direct recycling is seldom possible except for in-plant ("home") scrap. However, the variations between one plastic and another are greater than for the rubbers, and they are seldom concentrated in the way that tires pile up at service stations and auto-wreckers.

Plastics presently amount to between one and five percent of our municipal wastes. The range is wide because of the impact of, for instance, soft drinks and milk being marketed in plastic bottles in one area and not in another. It would seem that only if fully automatic machinery is developed will a substantial proportion of domestically discarded plastics be sorted for reuse.

If the plastics as such are separated, it is not clear at present whether or not further sorting into various types of plastics will be worthwhile. Thermoplastics – those which soften on being heated – can be reused, with some degradation of properties, by injection molding, casting, or other form of hot working, probably similar to the original method of processing. Thermosetting plastics, such as the phenolics, do not soften with heat and can be reused only by grinding and using as a filler in other plastic formulations [48].

LEGISLATION ENCOURAGING RECLAMATION

At present, reclaimed materials have to compete against so-called "virgin" materials with two economic penalties. The first is that the reclaimed material has first to be collected and sorted from a multitude of contaminants, in the way that a junk car must be collected from a country roadside and then must be separated into all the components which man has ingeniously combined to make an automobile. In contrast, iron ore, for example, can be mined in huge quantities in one place, and the sorting-separation process to win the iron from the ore can be highly mechanized.

The second economic disadvantage is that under Western laws of property, someone buying land normally owns everything beneath the surface of that land and can remove it. Since this is a virtually "new" concept, from the historical viewpoint, there does not seem anything particularly wrong with it to our myopic vision. But if the Romans or the Babylonians or the ancient Chinese had taken their civilizations a little further, we might now be lamenting that in those days there were bountiful sources of raw materials called oil and coal and gas which were all taken out of the ground and burned; and the copper and manganese and tin were found in rich ores which were so easy to remove and extract that they were all exploited in the course of a mere century or two. This is the legacy which we are reserving for our children.

We desperately need a new law which would provide a tax fund for the future. The tax would apply to all materials, whether or not they be considered to be resources, which are removed from the earth, water or air, and are not replaced by natural or man-induced processes. The tax rates would be different for different materials. The uranium tax might be $200 a ton, the copper tax, $50 per ton, and so on. Just as important, there would be a tax, say one dollar per cubic yard or per ton, on the removal of overburden or soil unless it were replaced in substantially the same form as it previously existed. There would therefore be a powerful economic inducement to reclaim materials, and therefore to design goods in the first place so that reclamation is easy; and in addition there would be an incentive to restore land after mining operations to as near as possible its original condition.

Reclamation Subsidies

Another area where present economics seem to discourage the public benefactor and encourage the despoiler is when the reclaimed materials do not find a ready market. The reasons for the lack of the market might be connected with the argument given above – that it is cheaper to extract virgin material when it is virtually free than to collect and transport and sort out contaminated materials. However, another economic distortion can enter the picture. It is the distortion of externalities. A graphic case is that of a compost plant, to be discussed more fully in the next chapter. The entrepreneur who designs, builds, and operates a compost plant brings large benefits to the community. Instead of dumping city trash and garbage where it will have a strong likelihood of forming breeding places for rats and insects and of polluting the ground water, the entrepreneur, at great difficulty and expense, sorts out the inorganic from the organic materials and produces something that can benefit, not pollute, the soil. But the value to the public is principally in the absence of pollution rather than in the presence of compost, and the price the public is willing to pay for compost does not usually cover the production cost. But the public should pay something for the absence of pollution, the external benefit the public receives from a compost operation. In other words, the compost operation should be subsidized to the extent of its public benefit.

In some countries, this subsidy is arranged by simply having the compost plant run by the municipality. In this country, we prefer to use private enterprise where possible because by so doing we can maintain incentives on productivity. In the salvage of paper, there is a very simple way of using private enterprise to the physical benefit of the public. In Cambridge, Massachusetts, and in many other cities, there are several entrepreneurs who drive the streets in pickup trucks and station wagons in advance of the city refuse-collection trucks. Their goal is to collect clean newspaper.

Operating to apparently random schedules and routes, they find it worth their while to collect newspaper and to sell it to the dealers at, currently, between six and fourteen dollars per ton. They are regarded as scavengers, and some communities pass laws against this practice. However, in Cambridge, the cost of collecting and disposing of all refuse is between \$30 and \$40 per ton, most of this being due to collection. The paper in the trash causes the greatest costs and nuisance in disposal. Accordingly, the benefits which the scavengers of newspaper are bringing to the city are in excess of the average \$30 to \$40 per ton. Of course, this benefit cannot be realized unless the collection of newspaper is scheduled and sufficiently substantial for the number of collection trucks and crews operated by the city to be reduced by at least one, although there is always some savings in that the cost of disposal, presently \$7.60 a ton at a commercial compactor plant, is reduced. Accordingly, the city has agreed to try an experiment. If this goes ahead, newspaper collectors will be invited to bid on a contract in a test area of the city. The city will either pay them a bonus of perhaps five dollars per ton over and above the income received from selling the newspapers, or will establish a support price to be made up by the city when the commercial price falls below it. The contractor will guarantee to pick up all newspapers which are set out on the curb in bundles on a designated day each week.

An important part of the proposed scheme is that a public-relations organization must be involved. The residents will not be required, but merely asked, to separate and bundle newspapers. The function of the public-relations firm is to encourage this salvage by means of advertising, through-the-door leaflets, distribution of special bags, prizes, and any other gimmicks which might have a payoff. It is possible that a separate contract may be made by the city with the public-relations firm whereby the firm would be paid a separate amount per ton collected over a certain minimum.

Another incentive towards reclamation is the "bounty" placed on the price of new cars. The arrangement usually proposed is that the bounty, which would be two or three cents per pound, to result in cars having minimum scrap values of \$50 to \$100, would be placed in escrow, or perhaps invested in improved scrapping plants, until the final owner either abandons his vehicle or turns it in for scrapping. In the former case the bounty would more than pay for the car's removal, and in the latter case the last owner would collect the bounty in cash.

For smaller items, bounties do not have much possibility of working, as has been found for returnable bottles. Accordingly, straight "disposal" taxes of a cent or two per pound on goods which have soon to be discarded such as containers or appliances have been suggested. These contain a beneficial incentive towards reduction of use of materials, but the incentive cannot be strong enough to have much effect. Also, there is no incentive towards making more reclaimable the goods and packaging which are involved. And there may be engendered a sentiment in the public that, as disposal has been paid for, refuse can be scattered in good conscience. Such a tax therefore seems to be possibly a good way to raise money, but not necessarily an incentive to saving resources or promoting reclamation.

REFERENCES

1. Waste Trade Journal, Charles H. Lipsett, Publisher, Atlas Publishing Company, New York, New York.
2. Fiber Market News, Market News Publishing Corporation, 156 Fifth Avenue, New York, New York 10010.
3. Paper Stock Institute of America, a commodity division of National Association of Secondary Material Industries, Inc., "Paper Stock Standards and Practices" Circular PS-66, effective as of January 1, 1966, issued from Association Headquarters, 330 Madison Avenue, New York, New York 10017.
4. Leonard Stefanelli, President, Sunset Scavenger Company, "Economics of Recycling Presents Bleak Picture" Solid Wastes Management/RRJ, January, 1970.
5. Frank Flintoff, "No Risk of Over-production in Growing Waste-paper Market" – *Municipal Engineering,* September 26, 1969.
6. W. J. Johnson, "Refuse-reduction Plant Montreal-Quebec" Engineering Journal, June, 1969.
7. The Institution of Mechanical Engineers, 1 Birdcage Walk, Westminster, London, S.W. 1, conference on "British Postal Engineering" London, 19th-21st May, 1970, published in Proceedings, 1969-70, Volume 184, Part 3H.
8. M. L. McKenna, "Sorting Technology and its Application to Materials Management," ASME paper 69-MH-14, 1969.
9. K. S. Goldstein, "Telecar Baggage System – Automation in Action" ASME paper 69-WA/MH-2, 1969.
10. Robert F. Vokes, "Extraction System for Paper-fiber Reclamation from Municipal Wastes," National Industrial Solid-Wastes Management Conference, Houston, Texas, March, 1970.
11. Martin H. Stanczyk, John A. Ruppert, "Continuous Physical Beneficiation of Metals and Minerals Contained in Municipal Incinerator Residues" a paper in the *Proceedings of the Second Mineral Waste Utilization Symposium,* jointly sponsored by the U.S. Bureau of Mines and IIT Research Institute, Held in Chicago, Illinois on March 18-19, 1970.
12. S. L. Blum, "Recycling Man-made Ores for Industry," presented at the 1970 Engineering Foundations Research Conference on "National Materials Policy" Henniker, New Hampshire, July 16, 1970.
13. *Science Journal,* London, "Potatoes Harvested by X-rays," January 1970.
14. Richard A. Boettcher, "Air Classification for Reclamation of Solid Wastes" a paper in the *Proceedings of the Second Mineral Waste Utilization Symposium,* jointly sponsored by the U.S. Bureau of Mines and IIT Research Institute, held in Chicago, Illinois, on March 18-19, 1970.
15. E. Douglas, T. Walsh, and A. Whitehead, "The Development of Equipment for the Dry Concentration of Minerals," Warren Spring Laboratory, Herts, U.K., August, 1967.
16. David Gordon Wilson and Ora Everett Smith, "Mechanized Reclamation from Municipal Solid Waste" paper given in the National Industrial Solid-Wastes-Management Conference, Houston, Texas, March, 1970, proceedings to be published April, 1971.
17. Stephen D. Senturia, Frank Winkler, David Gordon Wilson, Kendall S. Lewis, Hubert Hibberd, "New Sensors for the Automatic Sorting of Municipal Solid Waste," to be presented at AICHE, Cincinnati, Ohio, May, 1971.
18. Charles H. Lipsett, "Waste Paper" Chapter 11, *In* Industrial Wastes and Salvage (second edition), 1963.
19. A. Porteous, "The Recovery of Industrial Ethanol from Paper in Waste" presented to a joint meeting of the Scottish Branch of the Institution of Chemical Engineers and the Aberdeen Branch of the Society of the Chemical Industry at the University of Aberdeen, May 8, 1969. Published in *Chemistry and Industry,* December, 1969.

20. R. D. Fagan, A. O. Converse, and H. E. Grethlein, "The Economic Analysis of the Acid Hydrolysis of Refuse" *In* Reuse and Recycle of Wastes – Proceedings of the Third Annual North Eastern Regional Antipollution Conference, University of Rhode Island, July 21-23, 1970.
21. Richard C. Corey, "Pyrolysis, Hydrogenation and Incineration of Municipal Refuse – A Progress Report," a paper in the Proceedings of the Second Mineral Waste Utilization Symposium jointly sponsored by the U.S. Bureau of Mines and IIT Research Institute held in Chicago, Illinois on March 18-19, 1970.
22. Herbert R. Appell, Irving Wender, and Ronald D. Miller. "Conversion of Municipal Refuse to Oil" *In* Reuse and Recycle of Wastes, Proceedings of the Third Annual North Eastern Regional Antipollution Conference, University of Rhode Island, July 21-23, 1970.
23. F. H. Miller, "Conversion of Organic Solid Wastes into Yeasts; an Economic Evaluation" Public Health Service Publication No. 1909. Washington, U.S. Government Printing Office, 1969, p. 173.
24. *Technology Review,* February, 1970, "Bacterial Cattle Feed from Garbage?" by W. Dexter Bellamy of General Electric, Schenectady Research and Development Center, presented to the American Chemical Society in September, 1969.
25. John M. Harkin, "Uses for Sawdust, Shavings and Waste Chips" *In* U.S. Department of Agriculture, Forest Service, Forest Products Laboratory, Madison, Wisconsin, U.S.D.A. Forest Service Research Note FPL – 0208, November, 1969.
26. D. G. Wilson (editor) "Summer Study on the Management of Solid Wastes" Final Report, Volume I, published by the Massachusetts Institute of Technology Urban Systems Laboratory, Cambridge, Massachusetts, September, 1968. (Volume II has not been released)
27. Ward R. Malisch, Delbert E. Day, and Bobby G. Wixson, "Use of Waste Glass for Urban Paving" a paper in the Proceedings of the Second Mineral Waste Utilization Symposium jointly sponsored by the U.S. Bureau of Mines and IIT Research Institute held in Chicago, Illinois on March 18-19, 1970.
28. M. E. Tyrrel, and I. L. Feld, "Structural Products made from High-silica Fractions of Municipal Incinerator Residues" a paper in the Proceedings of the Second Mineral Waste Utilization Symposium jointly sponsored by the U.S. Bureau of Mines and IIT Research Institute, held in Chicago, Illinois on March 18-19, 1970.
29. John H. Abrahams, Jr. "Utilization of Waste Glass" a paper in the Proceedings of the Second Mineral Waste Utilization Symposium jointly sponsored by the U.S. Bureau of Mines and ITT Research Institute held in Chicago, Illinois on March 18-19, 1970.
30. John Frank Elliott, information from a materials-science seminar at M.I.T., October 23, 1970.
31. *Waste Trade Journal,* "Competition by the Smaller Steel Mills who Use Scrap" fifth paragraph, Volume 66, No. 45, November 7, 1970 issue.
32. *Waste Trade Journal,* "Blames Scrap Prices for Poor Business" Page 18, Volume 66, No. 44, October 31, 1970 issue.
33. *Waste Trade Journal,* "Lukens Steel to Increase Pellet Use and Less Scrap" Page 12, Volume 66, No. 40, September 30, 1970 issue.
34. *Waste Trade Journal,* "Coreco Metal Separator" Page 15, Volume 67, No. 8, February 27, 1971 issue.
35. *Waste Trade Journal,* "Hard Times for Copper-rich Lands" Page 2, Volume 67, No. 11, March 20, 1971 issue.
36. Robert F. Testin, "Recycling of Used Aluminum Products" *In* Reuse and Recyle of Wastes, Proceedings of the Third Annual North Eastern Regional Antipollution Conference, University of Rhode Island, July 21-23, 1970.
37. *Waste Trade Journal,* "115 Million Aluminum Cans Collected for Reclamation" Volume 67, No. 4, Page 6, January 30, 1971 issue.
38. *Solid Wastes Management,* "Senate Bill would Prohibit Misuse of Ecology Theme," Pages 12 and 54, March, 1971, issue.
39. Gerard C. Gambs, "Power Plant Ash – the Coal Industry's Hidden Opportunity" paper presented at AIME Annual Meeting, New York, New York, February, 1966.

40. Gerard C. Gambs, "Expanding the Market for Fly Ash" *Mechanical Engineering,* January 1970 issue, Page 26.
41. Howard K. Eggleston, "The Successful Utilization of Iron and Steel Slags" a paper in the Proceedings of the Second Mineral Waste Utilization Symposium jointly sponsored by the U.S. Bureau of Mines and IIT Research Institute held in Chicago, Illinois on March 18-19, 1970.
42. John H. Faber, and Philip G. Meikle, "Ash Utilization Techniques Present and Future" a paper in the Proceedings of the Second Mineral Waste Utilization Symposium jointly sponsored by the U.S. Bureau of Mines and IIT Research Institute held in Chicago, Illinois on March 18-19, 1970.
43. *Waste Trade Journal,* "Slag Piles can be Reclaimed" Page 4, Volume 67, No. 10, March 13, 1971 issue.
44. Charles H. Lipsett, "Rubber Scrap" Chapter 52, *In* Industrial Wastes and Salvage (Second Edition) 1963.
45. *Waste Trade Journal,* "Recycling of Scrap Passenger Tires" by James J. Barnett, Page 8, Volume 67, No. 8, February 27, 1971 issue.
46. J. A. Beckman and J. R. Laman, "Destructive Distillation of Used Tires" *In* New Directions in Solid Wastes Processing, 1970, Proceedings on an Institute held at Framingham, Massachusetts, a publication of the Technical Guidance Center for Industrial Environmental Control University of Massachusetts, Amherst, held May 12-13, 1970.
47. *Waste Trade Journal,* "Old Tires and Old Bottles now have a Market" Page 13, Volume 66, No. 44, October 31, 1970 issue.
48. Charles H. Lipsett, "Plastic Scrap" Chapter 1, *In* Industrial Wastes and Salvage (Second Edition) 1963.

CHAPTER 9

COMPOSTING MUNICIPAL REFUSE

Compost is composed of the products of bacteriological breakdown of organic materials – garbage, leaves, paper and so forth – and finds extensive use in improving the properties of soils for agricultural purposes. Composting has been practiced for centuries; the earliest practitioners appear to have been farmers in Asia who buried organic farm wastes and, after several months, reclaimed a compost produced by anaerobic degradation. The unpleasant odors of methane and hydrogen sulfide which accompany this very slow process gradually created widespread dissatisfaction with the anaerobic process and led to the eventual development of the aerobic process. In aerobic fermentation, the refuse is constantly exposed to air and an exothermic process results due to the biological activity of the micro-organisms which inhabit the refuse.

Normal household refuse contains large numbers of micro-organisms – including bacteria, protozoa and fungi – and the vast variety of food substances necessary for the preservation and growth of these populations, [1]. When sufficient quantities of air are provided, the micro-organisms breathe in the oxygen, ingest large quantities of the surrounding organic material, exhale carbon dioxide, and release in the form of heat the energy of oxidation. Certain of the elements contained in the refuse are metabolized by the organisms – these are necessary for the rapid growth and reproduction of the population. The entire process is highly exothermic, and the material may reach temperatures in excess of 180F as a result of the intense biological activity. The net result is a biological breakdown of the complex compounds contained in refuse to simpler compounds which more readily contribute to the functions of soil during plant growth. By proper adjustment of composting conditions, the peak temperatures will be above 150F in all parts of the mass, and the bulk of the digestion reactions can be completed in a matter of days. The thermophilic bacteria essential to the process thrive at these temperatures; moreover these temperature levels are sufficient to destroy most, if not all, pathogenic bacteria and insect larvae. Subsequent "curing" operations complete the degradation into a relatively stable product, yet one which retains a high degree of biological activity.

By way of contrast, in the anaerobic process, no temperature rise occurs because of the extreme slowness of the process (of the order of months); thus pathogens and insects can survive and multiply. The anaerobic process is the dominant mechanism in normal landfill operations, whether of the "basic sanitary" type or not (Chapter 6). However, the term compost usually implies material which finds agricultural application at a point distant from its production. Thus, compost manufacture is carried out almost exclusively by aerobic processes which can be completed rapidly, and which generate no obnoxious odors.

Harding has provided the following excellent definition and description of compost [2]. "Finished compost may be designated by the general name 'humus.' Although the term humus is not subject to rigid definition, it may be described as a complex aggregate of amorphous substances resulting from microbiological activity in the breakdown of plant and animal residues. Chemically, humus is a heterogeneous material including various compounds synthesized by microorganisms, complexes resulting from decomposition, substances undergoing decomposition, and plant material resistant to further breakdown. The principal constituents may be classed as derivations of lignins, proteins, certain hemicelluloses, and cellulose.

"In general, humus or compost is characterized by certain features:

1. It is dark brown to black in color.
2. It is practically insoluble in water, although a part of it may go into colloidal suspension.
3. It dissolves to a large extent in dilute alkali solutions giving a colored extract. For this reason, many early observers mistakenly defined humus solely in terms of "humic acid."
4. It has a carbon-nitrogen ratio ranging from 10:1 to 20:1, depending on the original material and its degree of decomposition from which the humus was formed.
5. It is not in a biochemically static condition: it changes in composition, through the activities of microorganisms, as long as environmental conditions such as moisture and temperature are suitable, until it is eventually oxidized to inorganic salts, carbon dioxide and water.
6. It exhibits a high capacity for base exchange, for combining with other inorganic soil constituents, and for water absorption with consequent swelling.

"When used in the soil, compost has many characteristics beneficial both to the soil itself and to growing vegetation. Its nutritional values include nitrogen, phosphorus, and potassium, although not in amounts equal to those of commercial fertilizers. In addition, it contains many trace elements essential to optimum plant growth. In conjunction with commercial fertilizers, compost exhibits additional desirable characteristics. Organic acids resulting from the metabolic breakdown of organic material form a complex with the inorganic phosphate. In this form phosphorus is more readily available to higher plants. Both phosphorus and nitrogen are involved in a storing effect peculiar to humus materials. The precipitation of phosphorus by calcium is inhibited; and nitrogen is converted as the bacteria die and decompose, thus preventing leaching of soluble inorganic nitrogen and making its rate of availability more nearly equal to the rate at which plants need it. This decomposition of insoluble organic matter by microorganisms results in the continual liberation of nitrogen as ammonia, which is then oxidized to nitrites and nitrates."

HISTORICAL REVIEW

The early composting processes merely involved the haphazard massing of garbage and perhaps manure or night soil into a single heap. The resulting anaerobic process proved most unsatisfactory for any large-scale applications.

It is generally agreed that the principles of composting were first enunciated by Sir Albert Howard who developed his ideas working in the Indore State of India early in this century [4]. The so-called Indore process involved the stacking of alternate layers of straw or leaves and manure or other organic matter. The loosely gathered straw allowed air to permeate the mixture and resulted in aerobic degradation of the organic material. Aeration was enhanced by inverting the entire heap after five days and reinverting it after an additional fifteen days. After a total reaction time of one month the majority of the degradation had taken place and the material could be set aside to "cure." This approach is still followed, although mechanization of aeration and turning has greatly accelerated the process.

Whatever aeration process is used, the exact time for transfer from the "active" sites to the curing bins is somewhat arbitrary and will vary with the nature of the feed material and desired characteristics of the final product. In general, however, it can be said that transfer is appropriate when the reaction rate is sufficiently slow that the process for transferring oxygen to the reaction site within the compost provides no impediment to the overall reaction rate. In other words, the rate of biological activity is sufficiently slow that oxygen can easily diffuse or be convected from the surrounding air to the center of the curing heap.

After a curing period of several months, biological activity will have fallen to a level which allows the material to be termed "stabilized" and ready for use. Biological activity never actually ceases and residual activity, albeit very slow, may be crucial to the utility of compost in improving the growing characteristics of a soil.

The original Indore process was developed for stabilizing the waste products of agriculture [1]. However, its application spread worldwide and it was demonstrated that a variety of feed materials, including sewage sludge and household garbage, could readily be processed using the same technique.

Recent Trends

In the last few decades the process of composting has undergone rapid changes in popularity when viewed in comparison with alternative waste-management techniques. The inherent desire to "gain something for nothing" has, at various points in time, caused composting to be passed off on the public as a panacea for all disposal problems. In some instances, large facilities have been constructed without the proper analysis of process economics or an accurate determination of process conditions. The resulting failures then discredit all composting operations. However, failures are not inevitable.

It would appear that any major metropolitan waste-disposal system should consider the use of composting for at least a portion of the waste stream. Many past problems in applying the process can be traced directly to an "all-or-nothing" approach. Compost is but one of many products which can be reclaimed from municipal waste. In the future, both ecological and material conservation forces will necessitate increased reclamation and recycle of components from our waste streams. Compost should be regarded as a candidate for reclamation along with paper, steel, glass, aluminum, and textiles. The incentive for recovering each component will vary widely with the economic and industrial conditions of the particular geographical area under consideration. In the case of compost, the type of soil in the area and the extent and nature of agricultural operations in the surrounding regions are important factors in determining the viability of a compost operation.

Economies of Scale

As has been demonstrated previously, efficiency and economy in solid-waste management can be achieved only by assembling a large stream of solid waste. A flow rate of 500-1,000 tons per day appears to be a target range which provides reasonable processing economics without the necessity of long hauls from distant areas. This size range may require the grouping of a number of political units, not only to achieve economies of scale, but also to have the choice of a number of alternative or complementary waste-management techniques. For example, it might be desirable to construct a composting facility adjacent to a large landfill or incinerator but to produce compost only during spring and summer when agricultural demands are high. Alternatively, the compost may be produced solely for cover material for a sanitary landfill and for other municipal needs.

European Leadership

As in many other areas of solid-waste management, the Europeans have established leadership in composting processes. A recent U.S. Public Health Service survey of Europe [3] documents some current European operations, examines the economics of each facility and speculates on the applicability of European techniques in the United States. The research activities carried out in conjunction with many of these facilities appear to be outstanding. Much has been done to elucidate the fundamental natures of the many biological processes which take place during composting and the process itself has been highly successful in certain geographical regions. For example the plant in Bad Kreuznach, Germany is located near the center of the Riesling wine area. In this region, many of the vineyards are on steep hillsides where large quantities of organic humus are required for proper plant growth and for the prevention of soil erosion. In this city, the compost facility is actually owned by a farmers' cooperative and local cities pay a fee to dump their refuse at the plant. Here, as in other regions of Europe, a successful plant is operated because there is a strong economic incentive to do so.

However, the economics of European operations do not inspire confidence that composting can be carried out on a large scale in this country. Analogous regions exist in the United States but a variety of political, economic, and social reasons have thus far prevented any truly successful operations.

Modern Refinements

Present-day composting operations have descended with remarkably little alteration from the earliest manual processes. Modern refinements have resulted primarily from more sophisticated means of achieving aeration; both forced convection of air (sometimes enriched with oxygen) through the pile and mechanical movement of the compost have been attempted in various equipment configurations. Significant effort also has been expended in developing grinding methods to reduce the particle size of the compost; the resulting increased surface area greatly speeds up the overall reaction rate and reduces the required time for primary composting and the secondary curing operation. The resulting improvements in capital and operating expenses have been significant.

More recently, substantial efforts have been directed at the removal of undesirable components from the waste stream prior to composting. Some commercial practitioners of composting have found that the potential values in some of these component streams far overshadow the worth of the compost itself, [4]. Many symbiotic relationships exist between reclamation and composting; and when viewed in proper perspective, compost should be considered merely as one of the components eligible for removal from a waste stream.

In addition to development work on the unit operations of composting (grinding, sorting, reacting, etc.) there have been significant contributions to our understanding of the fundamental processes which occur during composting. The most important variables of interest, other than the composition and particle size of the feed material, appear to be the moisture content and pH of the feed, the rate of oxygen feed, and the temperature of reaction [5, 6, 7].

Chemical Composition of Municipal Refuse and Compost Products

The composition of municipal wastes varies widely with geographic region, but more importantly, composition is highly variable with season even within a single area. To some degree, the composition of a feed can be changed by adding variable amounts of other materials such as sewage sludge, and by removing materials such as paper and glass, but even so, the overall composition will vary enormously with time.

The carbon-to-nitrogen (C/N) ratio is frequently used to measure the utility of compost in agricultural applications. Fortunately most organisms responsible for composting reactions are active over a wide range of C/N ratios. These organisms consume carbon and nitrogen in a ratio of approximately 30:1 in producing the proteins necessary for the growth and reproduction of the active organisms. Actually only about one-third of the reacted carbon enters the composition of the cellular protoplasm of the new organisms. The other two thirds are oxidized to carbon dioxide in the respiration process of the organisms [3]. Thus the C/N ratio decreases steadily during the digestion process.

The maximum C/N ratio in the fresh refuse should be about 30:1. A higher-ratio feed can be decomposed only partially because the low nitrogen value limits the extent of protein production and of organism growth. Actually when all the nitrogen is consumed, the organisms start to die and their nitrogen content is consumed by the organisms which remain. Thus the process can continue, but at a much reduced rate. A C/N ratio of less than 30:1 in the feed is both acceptable and desirable, but usually cannot be achieved without synthetic enrichment of the mixture. With a C/N ratio in the feed of 30:1, a ratio in the fresh product from the reaction unit may have a ratio of about 23:1, and that of the cured product would be 20:1 or less. A maximum ratio of 20:1 is a requirement if the compost is to be applied to land without the benefit of synthetic sources of fixed nitrogen. If ratios in excess of 20:1 are used, soil micro-organisms would consume nitrogen already present in the soil during the process of consuming the excess carbon from the compost; thus the nitrogen required for the growth of plants would not be available in suitable quantities.

Economics of Enrichment

The table below shows the composition of fresh and cured compost as produced in the Bad Kreuznach process [11]:

	Percent in fresh compost	Percent in cured compost
Organic content	28.4	19.1
Nitrogen	0.48	0.46
Phosphorus (as P_2O_5)	0.29	0.31
Potassium (as K_2O)	0.41	0.45
Calcium (as CaO)	2.91	3.78
Magnesium (as MgO)	0.90	1.05

Kehren and Vaillant [3] give the following as the chemical analysis of a "typical" compost:

		Wt on a Dry Basis		
Nitrogen	(N)	0.4	–	1.5%
Phosphorus	(P_2O_5)	0.2	–	0.8%
Potassium	(K_2O)	0.4	–	1.2%
Sulfur	(S)	0.1	–	0.5%
Calcium	(CaO)	2.0	–	8.0%
Magnesium	(MgO)	0.2	–	0.8%
Sodium	(Na)	0.2	–	0.5%
Iron	(Fe)	0.2	–	1.0%
Copper	(Cu)	0.001	–	0.08%
Manganese	(Mn)	0.002	–	0.05%
Zinc	(Zn)	0.001	–	0.04%
Boron	(B)	0.001	–	0.05%
Molybdenum	(Mo)	0.0001	–	0.005%
Cobalt	(Co)	0.0001	–	0.002%

Even with all the concern over the nitrogen content of composts, the above tables show that the total nutrient content of compost is almost negligible when measured in economic terms. Keeping in mind that chemical nitrogen can be purchased today for less than eight cents per pound, that phosphorus is available for 15 cents per pound, and potassium for five cents per pound, the total nutrient value is only about 0.1 cent per pound of compost. Stated in an alternative fashion, substantial enhancement of the nutritive value of compost can be achieved by enrichment with chemical products in amounts which increase the cost of the compost by only a small amount.

Because of the low nutrient value in compost and the widespread availability of chemical fertilizers, agriculturists look to compost primarily for the physical properties which it can impart to a soil. Maximum enhancement of plant growth can be achieved by controlling the active organic content and the particle size. An active organic content of at least 15 percent is required by authorities in Israel [6] although the minimum requirement obviously varies widely with the particular crop to be treated and the type of soil available.

Particle Size

As mentioned previously, most modern composting operations reduce the particle size of the feed material in order to increase reaction rates and to yield a compost product with the desired texture. A proper particle size of the foreign matter (stones, glass, etc.) is also very important in the final application of the compost product. A maximum particle size of one inch seems to be required in order to guarantee a reasonably homogeneous mixture for general agricultural applications. A maximum size of perhaps 1/4 inch is a useful limit when preparing composts for greenhouse or nursery soils [6]. On the other hand, care must be exercised to minimize the amount of fines and dust, which are not helpful in improving soil consistency. Particle-size reduction can be accomplished in a variety of ways; these will be described in subsequent sections where specific processes are discussed.

Moisture Content and pH of the Feed Material

The activity of microorganisms responsible for composting reactions requires significant levels of moisture in the feed. This is because the active microorganisms both respire and ingest their food through delicate cellular membranes. Moisture is necessary for the maintenance of these membranes as well as for the dissolution and transport of food stuffs to the membrane surface. The ideal moisture content has been estimated at about 55 percent by mass with a minimum level of 30 percent required to sustain substantial biological activity [3]. Thus if the moisture content of the feed is less than 50 percent, water is usually added to the refuse. The moisture content should also be monitored and adjusted during the fermentation process since the high temperatures achieved during digestion tend to evaporate water and dry out the reaction mixture. Thus, most composting systems incorporate sprinklers.

There is also a maximum level of water which can be permitted in the feed. It has been found that a moisture content in excess of 60-65 percent produces a "mushy" mixture and severely inhibits the transport of air to the interior of the refuse [3]. Thus, during particularly rainy periods it is desirable to drain and dry the mixture or to blend the feed material with sufficient quantities of dried compost product.

The pH of raw refuse is generally quite acceptable for the promotion of composting reactions. Alterations in pH due to the addition of sewage sludge do not seem to affect composting rates although little fundamental work has been done in this respect. Only in the extremes of highly acidic or highly basic feeds is a pH adjustment required.

Reaction Temperature

As mentioned previously, the aerobic microorganisms present in all refuse operate most effectively at temperatures in the range of 140F-160F. These temperatures are created naturally by the metabolic energy liberated by the microorganisms when composting is carried out under ambient conditions. At these temperature levels, the reactions can take place sufficiently rapidly to complete most of the degradation in a four-to-six-day period. The final stages of reaction are completed by storing the compost in curing piles over a period of months.

Another benefit usually cited for the high temperatures of the aerobic reaction is the destruction of pathogens and insect larvae. While this is certainly qualitatively true, it is not clear that a complete kill is achieved in the four-to-six-day residence time in the primary reactor. This is a subject of current research [18, 11]. Care in the specification and maintenance of minimum temperature levels is particularly important in northern areas where winter temperatures can cause cooling of the reaction mass. Little difficulty should be encountered in this respect; however, care must be taken in the initial process design because a loss of temperature can lead to anaerobic conditions and severe operating problems. Temperature is most easily controlled by adjusting the temperature of the air fed to the system and by limiting the temperatures in the area of composting activity.

Rate of Air Circulation

The single most important factor in designing a compost process is the type and rate of air flow through the reaction mass. Only the presence of air can ensure the advantages of an aerobic, high-temperature fermentation. The flow pattern of the air must be sufficient to deliver adequate oxygen to the interior of the refuse and to provide for the removal of the carbon dioxide exhausted by the microbes. Obviously it is also desirable to limit the power requirements for circulating the air and to provide for temperature control of the air fed to the digestion chamber. There seems to be little fundamental information available on the air flow rate required to meet all these ends. However, Kehren and Vaillant [3] have determined empirically that approximately four cubic meters per ton per day are necessary to supply enough oxygen and remove the carbon dioxide.

There is a variety of ways of satisfying all of the requirements for an air-circulation system. In fact it is the manner in which air circulation is achieved that fundamentally differentiates one composting process from another. If the reaction mass is subjected to absolutely no mechanical movement and no mechanically produced air flow, about 18-20 days are required to complete the bulk of the fermentation. This time can be reduced to four-to-six days by providing for forced-air circulation and, mechanical agitation of the refuse.

Summary Remarks

The design of a modern reclamation facility requires careful specification of what materials are to be salvaged from the refuse stream. Various metals, glass and paper, as well as compost, may be economically salvaged; the selection depends most strongly on geographical location. If composting is to be incorporated into the system, careful attention must be paid to the composition and particle size of the product, to the moisture content and pH of the feed, to the temperature in the reaction mass, and particularly to the air-feed rate. A variety of compost processes has been developed over the years; many of these are named after the manufacturer, the geographical site of the plant, or the process inventor. Despite the confusion of the 20 or more names for these processes, each process must take into account all the aforementioned factors as well as factors peculiar to a geographical location. As a result, many of the processes are very similar and it is relatively straight-forward to classify them. The next section describes a few of these and discusses some of their relative advantages and disadvantages.

COMPOST RECLAMATION PROCESSES

Most aerobic compost processes require some preparation of the feed, some alteration of the feed composition – either chemically or with water – some means of delivering air to the feed, and a storage area for the final curing. Fundamentally these processes differ from one another only in the mechanical handling of the compost during the primary reaction period and in the way in which air is delivered to the reacting refuse. Thus compost processes can be conveniently classified according to the mechanical technique employed as shown on the table below.

1. Open piling or "windrowing"
2. Processing in mechanical plants with
    a. continuous movement
    b. stationary composting with intermittent turning.

Windrowing

As mentioned earlier, the "original" composting process developed by Howard and his colleagues in India involved stacking the refuse in trenches with intermittent manual turning. This process has evolved into the so-called windrowing process wherein the refuse is piled in rows (above ground) which are six-to-ten feet high and frequently over 100-feet long.

In this as in most other processes, the incoming raw refuse is usually passed first through a magnetic separator to remove iron and then hand picked to salvage valuable or noncompostible bulk items. This is followed by a grinding or shredding operation to homogenize the remaining refuse and to enhance reaction rates. The Dorr-Oliver rasper has achieved significant use in the shredding operation, although a wide variety of hammer mills also has been used for this application.

The composting which takes place in a windrow is very slow and often takes four-to-six months to complete. Because air is not forced through the refuse, the operator must constantly be on the alert for the onset of anaerobic conditions. Temperatures do rise in the windrows but neither the level nor the duration of the temperature rise is sufficient to eliminate pathogens or insect larvae.

In order to aerate as thoroughly as possible, the windrows are inverted two or three times during the long composting process. Most often, a standard front-end loader is used for this purpose; however, the Roto-Shredder machine [13] developed by Imco Corporation, has recently been introduced and shows a good deal of promise in improving composting operations (Chapter 6). This vehicular device straddles and moves along a windrow, leaving behind a shredded and well-mixed product. Initial tests by the manufacturer indicate that composting in a windrow system can be vastly improved by this unit; but the economic decision to utilize such a device would vary substantially with the type of composting process being used and with geographical region. This unit would have to compete with more mechanized composting processes if a new plant were being considered. However, its real market may be in those smaller communities where windrowing is now practiced and where improvements are sought in current operations.

Another variation on the general process of windrowing is the so-called Van Maanen system. In this process the windrows are inverted periodically by large "grab" cranes. Another Dutch process, that practiced at Arnhem, uses a pivoted-belt conveyor to distribute refuse in a series of concentric, circular windrows. Both processes are considered highly successful. The excellent engineering work in the design and construction plays an important part in this success. Other factors are the highly developed uses for compost and apparently a keen interest among citizens in using a process which has ecological value.

Mechanical Plants – Continuous Movement – Dano Process

The most widely used process in the world is the Dano process – named after the European company which pioneered its development in the early 1950's. Over fifty of these units are used throughout the world, and the Dano organization has been responsible for the popularization of composting as a viable solid-waste-management technique. As a first step in the Dano process, iron materials are removed magnetically. The refuse is then conveyed to a large rotating drum, called a Bio-Stabilizer. This unit, resembling the lime kilns employed in the heavy chemical industry, is shown in Figure 9.1. A typical unit may be 10-12 feet in diameter and over 100 feet in length. It is approximately three-quarters filled with refuse; since it is pitched slightly from the horizontal, the refuse moves slowly through the drum. The drum is rotated at approximately 1 rpm – this is adjusted to provide a mean residence time of approximately three days. Temperatures of about 125-130F are maintained within the drum. Air and water are supplied at varying rates depending upon the type of refuse being treated, and in some installations sewage sludge is added in order to improve the nutrient content of final product. The constant movement of the refuse within the drum provides excellent transport of the air to the microorganisms, and the constant abrading of the refuse also provides

a. The "Bio-Stabilizer" drum

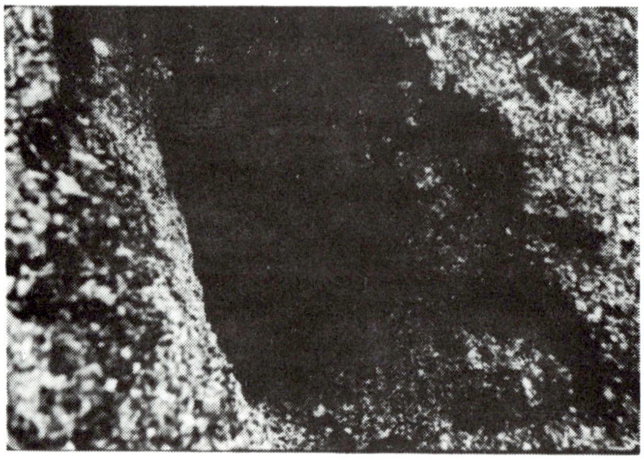

b. The product, showing whitish active region

Figure 9.1 Dano plant at Leatherhead, U.K.

excellent homogenization. In some cases acceptable particle-size ranges can be achieved solely by the Bio-Stabilizer; however, frequently a preliminary grinding step is needed to achieve optimum operation of the drum. As with many composting processes, the stabilized material leaving the drum is usually passed through final grinding and sieving operations to give a more uniform compost.

Most Dano installations are in Europe; they have been used successfully in Scotland, England, Germany and Switzerland. One or two plants have been built in the United States – these have been technically operable but financially unsuccessful. These financial problems, which seem to affect all composting processes in this country, resulted from poor financial planning and an underestimation of the problems to be faced in marketing compost and other reclaimed products.

Fairfield-Hardy

A second method employing horizontal movement of the compost is the Fairfield-Hardy process. This process has become particularly well known in the United States because of its successful application in Altoona, Pennsylvania, by the Fairfield Engineering Company. For many years, the city of Altoona had utilized a successful composting process which was based on a variation of classical windrowing. The city became particularly well known because its citizenry became accustomed to segregating its garbage (organics) from its non-compostable refuse. This provides a relatively uniform and attractive raw-material stream and eliminates the need for costly manual sorting in the plant.

In the early part of the last decade the process was improved substantially by the installation of the so-called Hardy digester. A schematic diagram of this unit is shown in Figure 9.2. The raw refuse is distributed evenly about the periphery of the cylindrical digester by a conveyor mounted on a bridge which rotates over the upper surface of the unit. The composting material is agitated by an array of rotating augers which are suspended vertically from the bridge. Air is fed constantly through a set of nozzles embedded in the foundation of the

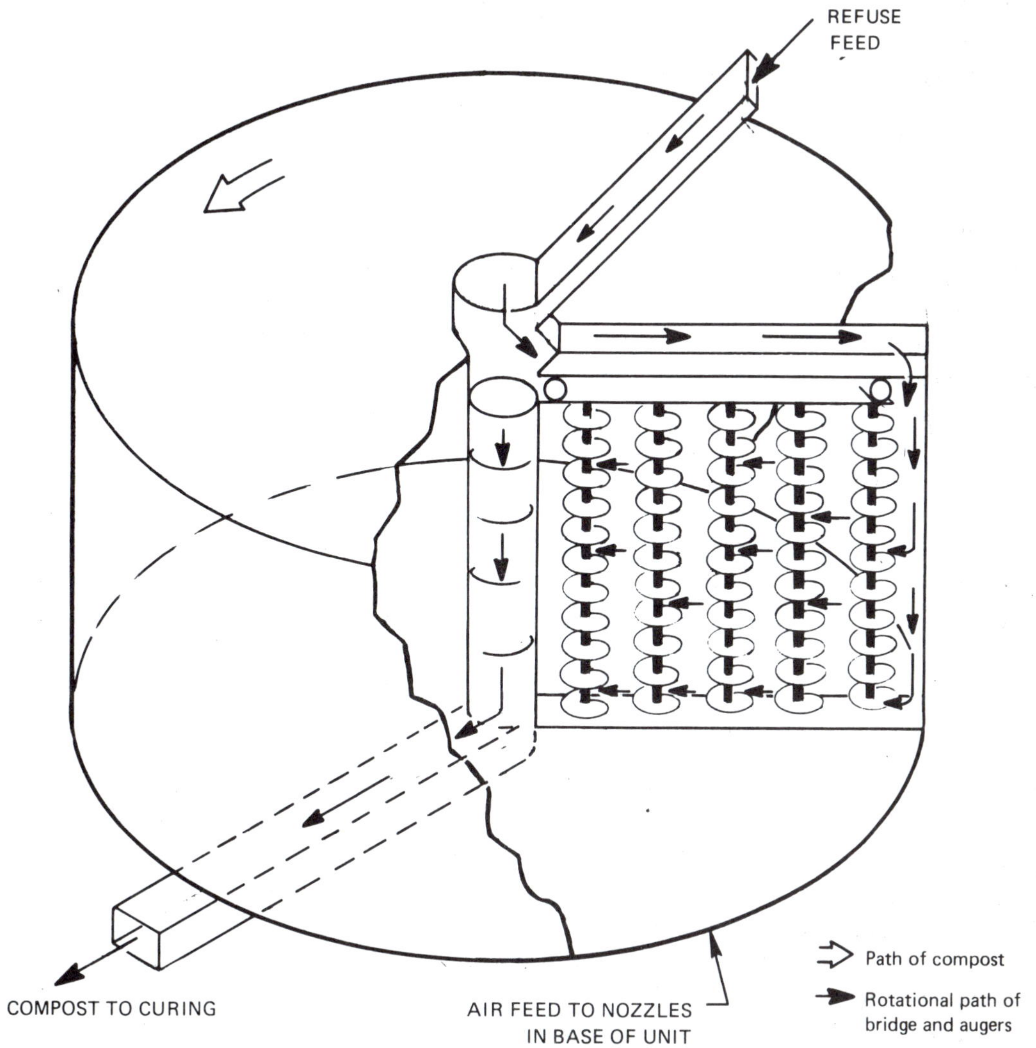

Figure 9.2 Diagram of hardy digester.

digester. The composting material is moved radially inwards by the action of the rotating augers of the digester. After a residence time of approximately five days, the stabilized product is removed through the hollow stand pipe located at the axis of the cylinder.

The most attractive feature of this unit is the intense mixing which it provides on a small scale. This ensures intimate contact of air with the refuse and a highly homogeneous product with excellent physical properties. One must pay a price for these features; a major component in that price is undoubtedly the maintenance of the unit itself. The mechanical mechanism is relatively complex and care must be taken to maintain maximum efficiency.

It has been estimated that Altoona F.A.M. Inc. processes approximately 6,500 tons of waste per year and receives a flat fee of $32,000 from the city [14]. If these estimates are correct, the fee works out to $4.92 per ton. Most of the compost product is sold in bulk to fertilizer blending plants in the area, and the company reports a very strong demand for its product.

The Fairfield-Hardy plant is often cited as one which should be emulated in other sections of the country. There is undoubtedly merit in this notion, but one must be very cautious in translating the process economics and product quality to other cities. This caution is particularly warranted because of the great receptivity of the Altoona citizenry in segregating its trash. Most knowledgeable observers are highly skeptical that such cooperation can be elicited from a significant fraction of the American urban public. Nevertheless the great success of the Altoona project deserves further scrutiny with the hope that at least some aspects of the operation can be transferred to other communities.

Earp-Thomas

A third general type of mechanical plant which continuously moves the compost is that which might be described as a vertical-movement design. Nearly all facilities which continuously move the compost vertically employ a set of horizontally deployed perforated trays which are spaced on top of one another as shown in Figure 9.3. Usually a four-to-six-foot clearance is provided between each pair of trays. In most designs, refuse is held and agitated on one tray for several hours and is then fed to the next tray below. Thus the refuse continually passes downward from one tray to the next, while warm air is provided continuously by a blower at the bottom of the cylindrical outer shell. The perforations on the trays and the continuous mechanical motion of the rakes provide for high oxygen-transfer rates and a high-quality final product. This general type of design is very similar to the classical Herreschoff furnace which has been used for many decades in the chemical and metallurgical industries.

The most widely known design of this type is the Earp-Thomas Multi-Bactor digester; a schematic drawing of this unit is shown in Figure 9.3. This process is employed on a worldwide basis, including a unit in Seoul, Korea; however, the most successful operations of this type appear to be in Heidelberg, Germany and Verona, Italy. Most applications of the Earp-Thomas process require an initial rasping and sieving, so as to enhance the composting rate and to minimize wear and jamming of the mechanical rakes. It has been reported that a maximum particle size of about one inch is necessary for efficient rake operation.

The ground and screened raw refuse is transported to the top of the compost tower as shown in Figure 9.3. In the Heidelberg unit, the tower is 20 feet in diameter and 35-feet high. The refuse passes serially through the tower's eight stages with a total residence time of approximately three days. In this operation the plows normally circulate only eight hours per day, but this is apparently sufficient to produce a final product which requires little or no cure

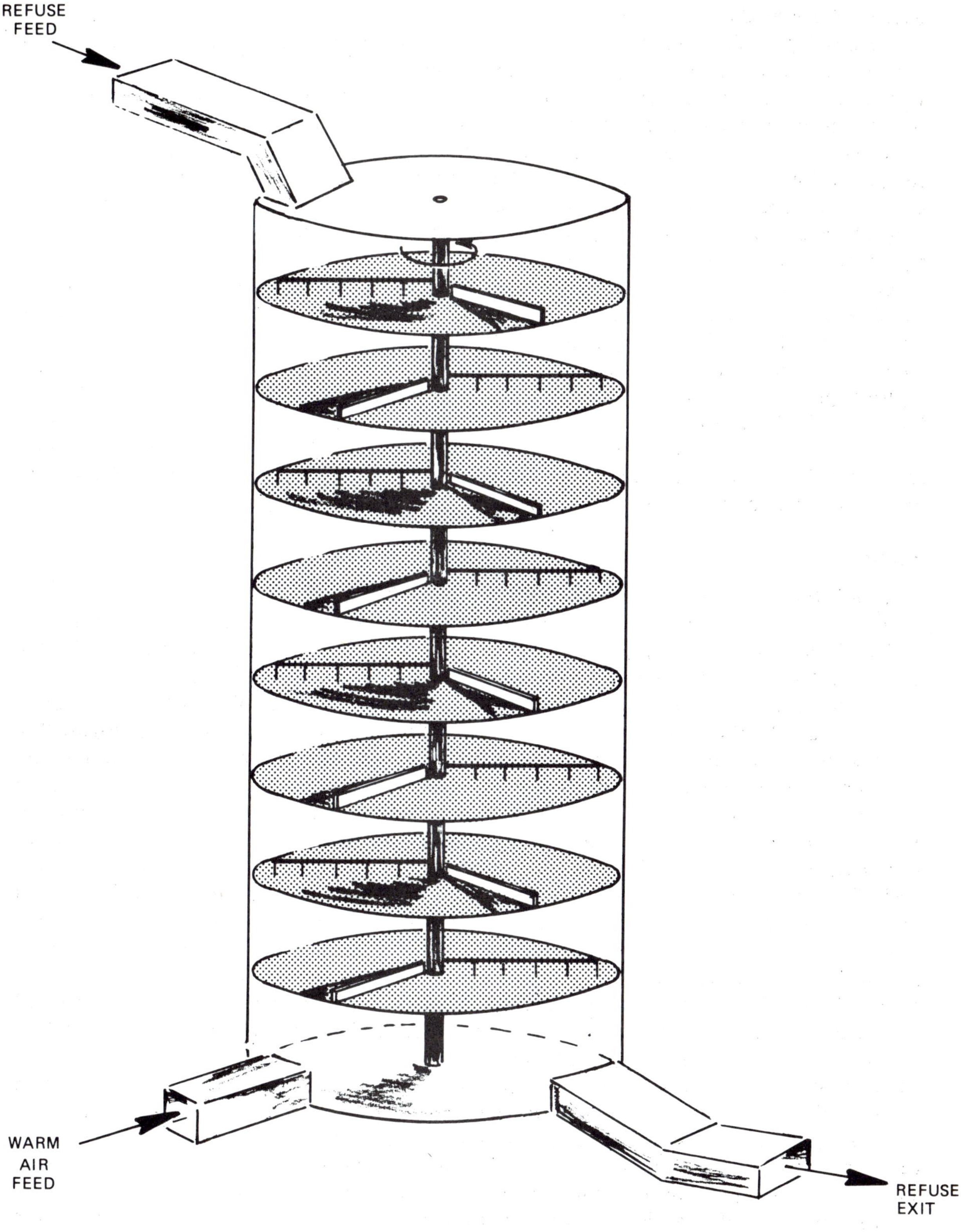

Figure 9.3 Earp-Thomas multi-bactor digester.

time. The product issuing from the tower is sieved again and fed to a vibrating separator where glass and other hard objects are removed.

The developer of the Earp-Thomas process has asserted that the success of the design is significantly enhanced by a proprietary type of bacteria which is inocculated into the refuse. This assertion has not yet been verified by an independent testing laboratory and it is generally agreed that those bacteria necessary for composting occur naturally in all normal refuse. This is particularly true since sewage sludge is frequently mixed with the refuse just as it is fed to the compost tower. Sewage not only enhances the moisture and organic content of the compost, but provides large quantities of bacteria for the process. Until demonstrated otherwise it is probably safest to attribute any success of the Earp-Thomas process to its mechanical design rather than to any unique biological processes.

As mentioned previously, there are many "vertical" processes employed throughout the world; but, in concept at least, most of these represent only minor departures from the Earp-Thomas design. For example the so-called Williamson process was employed during the 1950's to process four tons per day of refuse in Williamson, Michigan. The four-stage vertical digesters proved unsatisfactory since mechanical agitation was not provided in each stage and the only effective mixing occurred when the refuse was dropped from one stage to the next. The residence time averaged well over two weeks, and anaerobic conditions were frequently encountered. As a result this operation was discontinued after several years and after a variety of attempts to improve the plant operation.

Naturizer

A third type of continuous digestion system – the Naturizer System – represents a combination of features of the horizontal and vertical types of design. This design was originally utilized in a 150-ton-per-day plant in San Fernando, California which was developed by the Lockheed Aircraft Corporation; this plant is no longer in operation. Subsequently, a 100-ton-per-day plant based on the same design was opened by International Disposal Corporation in St. Petersburg, Florida. A schematic diagram of the St. Petersburg plant is shown in Figure 9.4.

As shown in this figure, the initial receiving, grinding, and manual-salvage operations are very similar to those applied in other systems. Paper and rags are removed manually and are baled for subsequent sale. A magnetic system employing both an overhead magnetic conveyor and a magnetic pulley beneath the refuse stream ensure the removal of virtually all ferrous materials. The prepared refuse stream is conveyed to the first of five insulated digestion "cells." Each cell contains a single conveyor belt which is approximately ten-feet wide and 150-feet long. The size of each conveyor is specified so that approximately one-day's feed (80-90 tons after salvage) can comfortably be distributed along its length. Thus the conveyor in the first cell moves slowly during the eight-to-ten hour work day and a day's feed is accumulated. All belts cease moving during the evening hours. The first three cells are arranged such that the refuse falling from the tail of one conveyor falls on to the head of the next. The mixing action achieved during the fall from one cell to the next provides the contact with the air necessary for the digestion process. After three days the partially digested refuse issues from the third cell and is fed to an intermediate grinder which provides new surface area for composting action to take place. The digestion process is completed in the fourth and fifth cells which are also arranged in a head-to-tail fashion. Temperatures of approximately 150F, adequate air flow, and proper humidity are automatically maintained in each cell.

After five days of digestion the product from the fifth cell is stored in a large heap and allowed to cure for a period of several weeks. After cure, the compost product may be shipped

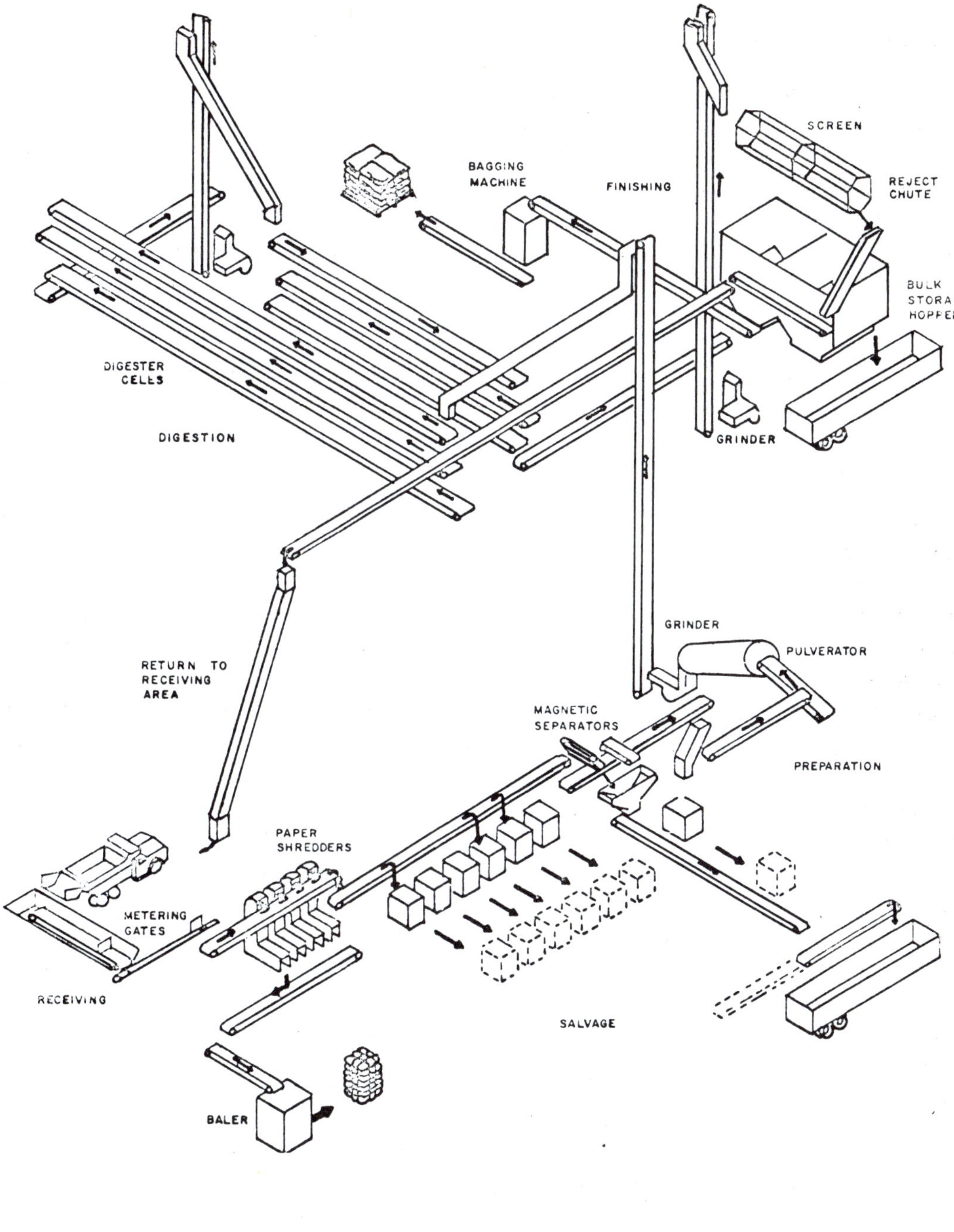

Figure 9.4 Diagram of the naturizer system.

directly in for bulk applications. Some of the compost is passed through a final grinding and sieving operation; this extra step is taken when application is intended for the nursery man or the home gardener.

Mechanical Plants – Stationary Processes

The continuous motion of the refuse given by the processes just described helps facilitate contact with the air necessary for aerobic processes and also assists in reducing particle size and improving the physical characteristics of the final product. Nevertheless, to one degree or another, all of these continuous processes are plagued by a common problem – the wearing and malfunction of mechanical components. For example, the conveyors used in the Naturizer system must operate at 150F and provide support and motion for over one hundred tons of material at any one time. Although the engineering design for these conditions can clearly be achieved, maintenance and component replacement are likely to be major concerns for production management and major costs when assessed by financial management. The malfunction of a critical mechanical component which causes only a few days of down-time not only can prove financially disastrous, but also can lead to severe operating problems.

Caspari-Brikollare

To counter the potential for mechanical failure, processes have been designed which eliminate or greatly reduce the physical movement of the refuse during the digestion process. One of the more successful of these is the Caspari-Brikollare process currently being utilized in Schweinfurt, Germany. As described by Hart [11], this process utilizes an initial magnetic separation and shredding in a Dorr Oliver rasper. Following the ballistic separation of such items as glass and stones, the shredded refuse is mixed with dewatered sewage sludge in an auger conveyor. The resulting material is fed to a proprietary machine which produces highly compressed briquettes, approximately 15 inches x 9 inches x 6 inches. These are stacked on pallets in a geometrical pattern which facilitates the flow of air during composting.

The heat produced by the aerobic digestion dries out the outer surface of each briquette. As the water evaporates, air enters the pores which are created and continues to act on the active receding surface within the briquette. Water is moved by capillary forces from the interior to the active surface. By proper selection of pressure during compression, proper specification of briquette dimensions, and proper stacking of the briquettes, it is apparently possible to achieve a balance of air and water transport within each briquette. Hart reports that aerobic conditions are maintained during composting and that a satisfactory product results. It is necessary to shred the briquettes prior to final distribution to the consumer.

Although the Caspari-Brikollare process achieves the desired end of eliminating mechanical motion of the refuse during digestion, it is far from clear that the process economics have been improved. The elimination of motion has been achieved at the expense of a rather complex briquetting machine. The incremental depreciation charges on this machine may well exceed the maintenance costs which have been eliminated. The process is quite new and little financial information is available. However, from a technical viewpoint it is clear the Caspari is extremely clever and represents one of the few real innovations in composting in many years.

Metro-Waste

Another attempt at curtailing mechanical problems during digestion has been aggressively pursued by Metropolitan Waste Conversion Corporation, Wheaton, Illinois. The Metro-Waste process, perhaps along with the Fairfield-Hardy process, represents the most successful application of composting yet achieved in the United States. However, as will be subsequently discussed, the Metropolitan Waste Conversion Corporation has encountered severe financial difficulties because of problems in marketing the reclaimed produce from its plants.

The company's first plant was built in Largo, Florida and since 1966 has successfully processed about 50 tons per day of municipal refuse from that city. The compost product has found markets with the local citrus growers as well as local nurseries and golf courses. A second Florida plant has more recently been constructed in Gainesville at a cost of $1,100,000. This 150-ton-per-day facility is jointly administered by Metro-Waste and technical staff from the University of Florida and has received financial support from the U.S. Public Health Service in the form of a demonstration grant.

The principal Metro-Waste activity is its 350-ton-per-day plant located in Houston, Texas, and operated by its subsidiary Lone Star Organics. This plant processes approximately 25 percent of Houston's refuse and a long-term contract with the city calls for a fee of, currently, about $4.25 per ton to be paid to the company. The plant is well located within the city limits of Houston, and ready access is available to that quadrant of the city allocated to this plant. The facility is very attractively designed as seen in the architect's sketch, Figure 9.5. The appealing design, together with an intelligent plant location within a partially industrial section, have minimized the type of problem with local residents which have plagued many other composting operations.

Figure 9.5 Houston compost plant — Metropolitan Waste-Conversion Corporation (Architects' drawing).

courtesy Metropolitan Waste-Conversion Corporation

A schematic diagram of the Metro-Waste system is shown in Figure 9.6. The process is quite similar to that described earlier for other systems. City trucks, after weighing in at the plant entrance, deliver refuse at a minimum rate of 300 tons per day to a sheltered conveyor system outside the plant building (at the far right in Figure 9.5). The oscillating conveyor shown in Figure 9.6 delivers the refuse to a manual salvage section located within the building. In this sorting section, laborers recover "clean" paper, glass and metal items; if current economic conditions permit, rags are also recovered and baled. Finally, the manual pickers also remove

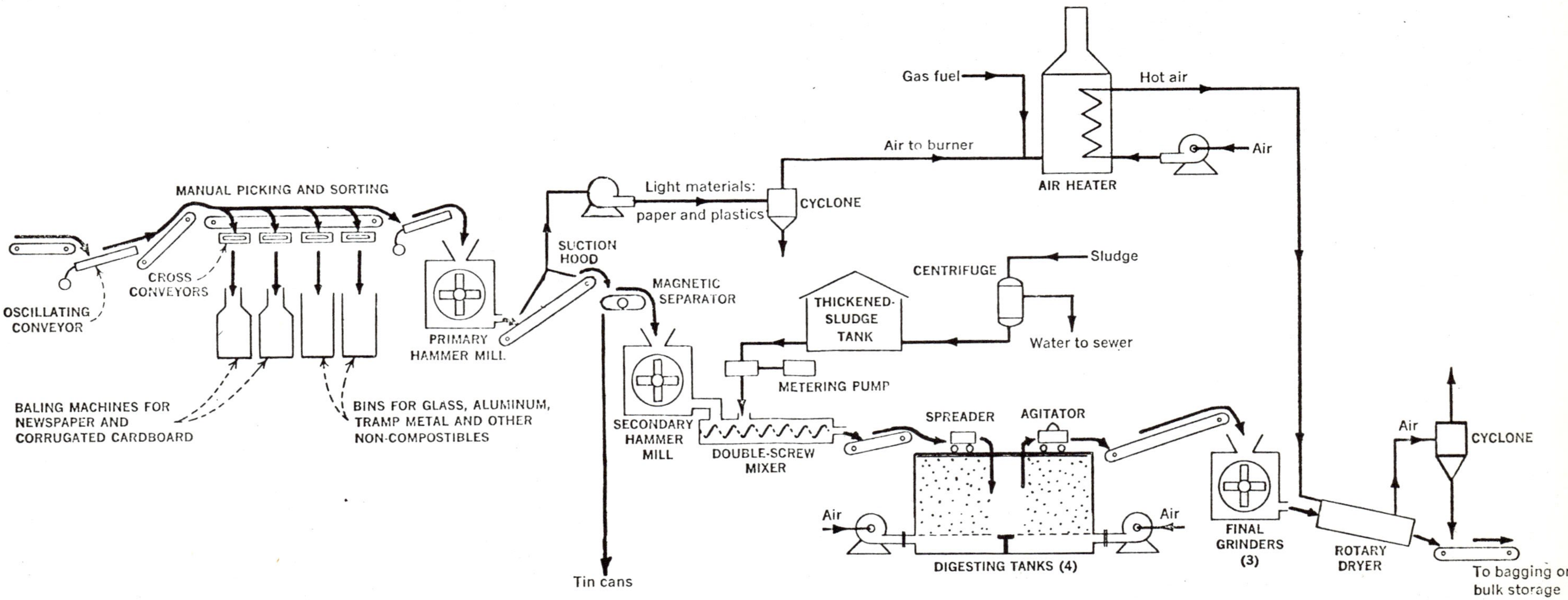

Figure 9.6 Diagram of the Metropolitan Waste-Conversion Corporation compost and salvage process [12].

bulky items such as rubber tires and bed springs, which are unsuitable for composting and potentially damaging to the hammer mills downstream.

The primary hammer mill reduces the remaining refuse to particles below a maximum dimension of one-to-two inches and facilitates the subsequent removal of light paper and plastics by vacuum and the magnetic recovery of tin cans and other ferrous items. The light papers and plastics are burned to produce process heat; building heat can also be supplied during winter months. The tin cans are deposited in rail cars (at the left in Figure 9.5) and are ultimately utilized in the recovery and refining of copper.

The main waste stream is given a second hammer-mill treatment, and provision is then made for adding sewage sludge prior to composting. The additional sewage sludge seemed to provide some processing benefits when it was tried soon after the plant start-up. However, occasionally it was found that the sewage sludge contained high concentrations of inorganic chemical wastes – apparently the products of improper industrial-waste disposal. In one instance, the high inorganic content completely destroyed the thermophilic bacteria essential to an aerobic process. The consequent unaerobic conditions caused operating problems which were very difficult to rectify. As a result, the process is now run without the inclusion of sewage sludge in the feed.

The refuse product of the second hammer mill now passes to the digesting tanks which embody the principal proprietary development achieved to date by Metropolitan Waste Conversion Corporation. The digestion and final processing is well described in the recent paper by Prescott [12].

> "Proper digestion is a key step in the process. At the Lone Star plant, the sludge-moistened compostable material is spread by Link-Belt trippers traveling on tracks into the digestion tanks. The Houston plant has four of these units, each measuring 360-feet long, 20-feet wide, and 8-feet deep. Each tank is capable of holding two days' production, and the normal residence time of the compost is six days.
>
> After the digester is half filled with the day's production, recording thermocouples are placed in the bed and aeration begins. While it proceeds, two blowers, each with a capacity of 50,000 cubic feet per minute, provide air to a plenum chamber under the compost bed. The air is easily directed by dampers.
>
> Inside the tanks, a balance between blowing too much air, which cools the material, and blowing too little, which would allow the metabolic heat of the bacteria to boost the temperature, must be kept at all times. At the Lone Star plant, the air flow is carefully timed to raise the temperature to 135F in the first 24 hours. At the end of the six-day cycle the bed's temperature is from 165-170F. Midway in the cycle the bed is turned over by a 15-ton agitator traveling on rails. This unit also empties the bed onto conveyor belts.[1] The records kept on each batch include temperature, air flow, moisture, and nitrogen, phosphorus and po-
>
> From the digesting tank the compost goes through three grinders in parallel; these reduce particle size to 0.5 inches. A rotary drier then cuts the moisture content in the compost to less than 20 percent, thereby stopping all bacterial action. Drying is necessary for bagged compost but can be bypassed for material shipped in bulk by rail or truck.

[1] at the end of digestion.

The processed compost is gray and has texture similar to that of peat moss. If allowed to bypass drying, the material gradually turns black and assumes the texture of a rich topsoil. Although used as a soil conditioner, the compost has an NPK ratio of 2-1-1, slightly better than steer manure.

The Lone Star plant operates eight hours per day for six days of the week only. Consumption of electricity runs to 200,000 kwh per month; that of gas fuel for the drier is about one million cubic feet per day. The plant has a master-control board that enables one man to monitor every processing step on closed-circuit TV."

The real issue of importance is Metro-Waste's demonstrated ability to provide a high-quality compost product while maintaining the refuse in a sanitary state during virtually the entire period of digestion. As Prescott mentions, the contents of each digester is inverted once during the process by a large agitator unit which also serves as an unloader. It is this device which is essential to Metro-Waste's proprietary process and serves as a basis on which to build further process improvements.

THE PRODUCTS OF COMPOST-RECLAMATION PLANTS – THEIR USES AND MARKETS

Overall Economics

An extensive survey of the markets in the solid-waste industry by the Carborundum Company [10] revealed that collection and transportation costs represent 60-85 percent of the total cost for disposal of municipal refuse. The collection costs range from $7.00 to $24.00 per ton with an average of approximately $15.00 per ton. Furthermore, Carborundum estimated that an average cost of $4.00 per ton is incurred in operating a sanitary landfill. In analyzing the economic viability of a composting process it is very helpful to keep these figures in mind. In contrast to landfill or incineration, composting is a unique disposal process in that virtually the entire stream of refuse must be transported both to and from the plant. Thus the overall economics of composting must account for double transportation and handling costs – costs which already dominate the financial analysis even when only a "one-way" delivery is involved. It follows directly that composting will succeed only when means are developed to sell the products of a compost-salvage operation – and to sell them profitably. Only when markets can be identified for paper, glass, recoverable metals, and compost can a compost-reclamation plant survive – even as a part of a larger waste-disposal complex.

The difficulty of managing a reclamation facility was highlighted in a Wall Street Journal article [9] which discussed the plight of the Lone-Star Organics plant in Houston. At that time the city paid the company $4.11 per ton to take 300 tons per day of its refuse.

Because of difficulty in selling the products of the plant, Victor Brown, the company president, estimated in the same article that they were losing $2.00 per ton of refuse processed. This figure is not surprising when one bears in mind the Carborundum estimate of $4.00 just to operate a sanitary landfill.[2] Houston quite clearly is receiving this service at a bargain price, while the company struggles to sell enough of its reclaimed products to solve its financial problems.

As the Journal [10] points out, reclamation and recycle are not new ideas, but they historically have been difficult to implement because of a wide variety of economic and

[2]Carborundum's estimate is higher than that of other authorities; see Chapter 6 – Editor.

sometimes political reasons. Each manufacturing industry has its own peculiar set of circumstances which tends to dictate the amount of recycling which takes place. For example, lead is used in a few, highly concentrated applications such as storage batteries and the ease of reclamation from these sources leads to a very high overall recovery – about half the annual production of is reclaimed and recycled. About half the iron is also recycled because of its concentrated use and ease of recovery magnetically. On the other hand, paper, which represents about half of all municipal refuse, is recycled only to a level of about 20 percent, with very little of this amount coming directly from the municipal-refuse stream. The Houston plant has had difficulty in producing waste paper of sufficiently high quality to be used by paper manufacturers in the production of lower-grade paper products. The only markets of any substance and consistency have been those which use waste paper in manufacturing building materials such as shingles and wall board.

Because of the vast amounts of paper available in municipal refuse, the ability to produce and sell waste paper is the single most important factor in managing a compost-reclamation facility. Compost plants are really paper-recovery plants which also produce other products such as tin cans, mixed glass cullet, rag bales, and compost. Nevertheless, the importance of compost and the markets for this material should not be underestimated, since compost is second only to paper in its impact on the economic viability of a reclamation facility. As a result, great effort has been expended to identify and develop markets for compost and associated products made from compost. Much of this work has been focused on: (1) the use of compost to replace peat moss in horticultural applications; (2) combining compost with synthetic fertilizers; and (3) hydromulching. A few remarks will now be made about each of these applications.

Compost as a Replacement for Peat Moss

Organic matter is an essential component in any soil. These materials not only control the physical properties of the soil such as degree of water retention and texture, but also are essential to the biochemical processes necessary for micro-organism activity and plant growth. For example, the organic component in a soil contains the fixed nitrogen, phosphorous and potassium which are required for plant growth and also serves as a food source for bacteria and other protozoa.

Peat moss is a widely used material for enhancing the organic content of soils. It is used both by the home gardener and by the commercial nursery man or landscape specialist. Significant amounts are also used by farmers for certain types of crops. Peat is a naturally occurring substance and is "mined" from bogs; many of the highest quality peats are found in northern climates where repeated freeze-thaw cycles provide an optimum consistency to the product. Peat is a rather bulky low-density material having a relatively low unit value, (as low as one cent per pound to the consumer). Thus shipping costs are a very important element in determining the ultimate price to the consumer. Nevertheless, some peats are of sufficient quality that substantial amounts are imported – most notable the German Sphagnum peats.

Various efforts are underway to force compost products into the markets now dominated by peat-moss suppliers and to develop new markets previously unavailable to the higher-cost peat products. First of all, markets can be developed in regions far removed from natural sources of peat moss. The high cost of shipping peat into the region gives a local compost producer a substantial economic advantage. For example, Lone Star Compost is attractively packaged in colorful, 50-lb. polyethylene bags and is sold under the trade name "Alive" to local home owners and nurserymen. A number of experiments have been run with local farmers by treating large tracts of commercial crops with Lone Star compost.

The markets for compost should be particularly large in the agricultural districts of the south. Many of the southern soils are very sandy and low in humus content. In addition the growing season is very long due to favorable weather conditions; thus it is difficult to build up the organic content by natural processes. Although compost is an ideal solution to this sort of problem, this market has been very slow in developing. Undoubtedly it is difficult to convince local farmers that compost can yield substantial economic benefits. Farmers are also particularly conscious of the nitrogen content in the fertilizers they use. Since an "average" compost will contain only one-percent nitrogen, it is difficult to convince farmers that the product has value. This is particularly true in the "farm-belt" region of the mid-west in which one finds soil very rich in humus content. Thus the reluctance of the farmer to commit a substantial initial cash payment for the compost has greatly restricted the agricultural market for compost. Even in the south, surprisingly few compost facilities have been built to serve local agricultural markets. Plants in Houston and Mobile are the exceptions to this observation, but their success in developing and serving their local markets is as yet unproven.

In addition to programs to serve local agricultural markets, efforts are afoot to merchandise compost to individual home owners through national retail chains. Retailers such as Sears, Roebuck and Company, Montgomery and Ward, and Kroger Stores already market substantial quantities of peat through their home-gardening subsidiaries. Some test marketing of compost products has been done through these types of retailers, but there appear to be some doubts that product quality will be sufficiently consistent to meet the demands of the consumer. In particular, questions have been raised about the appearance of the compost product and the potential that objectionable odors might develop [16].

One of the great advantages of compost is the wide variety of chemical elements it contains. Many such elements are necessary in trace amounts to foster plant growth. Elements such as boron, manganese, cobalt, molybdenum, and zinc are essential for the proper growth of vegetables. In addition, significant amounts of nitrogen, phosphorus, potassium, sulfur, magnesium and sodium are required for the growth of all types of plants.

While compost does not contain sufficient quantities of the "major" elements – nitrogen, potassium and phosphorus – to be called a fertilizer, it does, in contrast to peat moss, contain virtually all the necessary trace elements required for plant growth. This seems to be a little-appreciated fact, one which if properly understood could significantly increase the acceptance and use of compost by both the professional and amateur gardener.

The advantages of compost over peat moss in fostering certain types of plant growth have been demonstrated at Texas A & M University. This research, which was sponsored by Metropolitan Waste Conversion Corporation, compared the "Alive" compost peat with German Sphagnum peat which is generally recognized as the highest-quality material of its type [17]. The tests included the germination and growth of tomato and pepper seedlings as well as four varieties of coleus. Comparative tests were made by combining the peat materials with standard mixtures of sand, dolomite, gypsum and significant amounts of standard chemical fertilizers. In most cases, those plants grown with compost providing the organic content were fuller and larger than those grown in Sphagnum. This advantage may have been caused by the broader spectrum of trace elements available in the compost soil conditioner, but no work was done to substantiate this speculation.

In summary, it would appear on the surface that substantial markets should exist for a soil conditioner based on compost products. Compost-based conditioners appear to be at least as effective as the best peat mosses which are now successfully marketed nationally. Compost could also be sold at a lower price. However, despite intensive efforts, there have been no successful national marketing programs to date for compost-peat products. The Altoona –

FAM Company has a successful regional program in the Ohio-Pennsylvania area, and Hart [11] has reported in detail on successful regional programs in Europe where compost is used for erosion control in vineyard regions. There are undoubtedly a variety of regional markets which are yet to be successfully served in this country. The lack of success compost producers have had in replacing more than a tiny part of the peat-moss market indicates the need for a lengthy and costly national marketing program for compost. The national concern over the environment and in favor of recycling would greatly assist such a program.

Compost-fertilizer Combinations

Perhaps the most obvious way of improving the properties of compost is to increase its N-P-K content by adding powerful fertilizers. By increasing the content of the major elements required for plant growth, one overcomes the principal criticism of compost made by professional farmers. There are two basic ways in which the fertilizer value of compost can be enhanced; dry blending and chemical complexing.

Dry blending, as the name implies, merely involves the physical mixing of compost and a standard fertilizer compound. Often the entire mixture is pelletized so as to provide ease of handling at the plant and to facilitate mechanical application by the consumer. This general approach has been taken by the Altoona FAM Company which has been very successful in marketing most of their compost as a dry blend with a standard formula of nitrogen, phosphorus, and potassium. In fact it is claimed that the company could sell many times its current annual production of 2,200 tons of compost.

Dry blended products are successful in bringing the proper nutrients to the soil, but they suffer from an unsuitable rate of release common to many low-grade fertilizers. That is to say, plants require varying amounts of the different chemical elements during the growth cycle. The chemical components of a dry blend tend to be leached out by the first rainfall, thus providing the plant with a large initial stimulus. However, the fertilizer value may be entirely exhausted after this initial strong exposure.

This problem has been overcome by the premium-grade, slow-release fertilizers which have been introduced over the past few years. An example is Scott's Turfbuilder which releases its fertilizer chemicals slowly and in a manner consistent with the requirements of growing grass. An effort has been made by the Michigan Fertilizer Company to produce an analogous product. The patented process is based on peat moss and involves an initial exposure of the peat to ammonia in a high-pressure autoclave and then a final tumbling in an aqueous mixture of phosphoric acid and potassium chloride [16]. The final product contains not only the organics from the peat but also gives an analysis of 20 percent nitrogen, 10 percent phosphorus, and 5 percent potassium. Because these major elements are partially chemically bound to the peat, they are released slowly in a manner analogous to premium-grade fertilizers. This type of process probably could also be used with compost as a base, but it is not known if this has been tried.

Hydromulching

A third general class of markets for compost is frequently termed hydromulching. This process involves mixing compost, standard fertilizers, and grass seed in water to form a low-viscosity slurry. The slurry can then be sprayed onto a barren hillside or embankment; the resulting rapid plant growth is very effective in deterring soil erosion. This process has been used for some years, but on a limited basis, for replanting the embankments created during highway construction. For highway applications, the slurry can be transported in a water-tank truck and sprayed onto the areas with the most serious potential erosion problems. With the presence of a compost or peat-moss humus to provide organic content, the process has provided a low-cost grass cover on even the lowest-grade soils.

This process has shown a certain amount of merit especially in areas with very sandy soils. For example, some successful experiments were carried out in Los Angeles in the early 1960's using a slurry product called "Sta-Soil" manufactured by General Conversion Systems Corporation. Further applications of this general process have helped in the recovery of land which has been laid barren by strip mining. Other tests have been run in Florida under Federal sponsorship, but to date no significant nationwide market has been established.

THE FUTURE OF COMPOSTING-RECLAMATION PROCESSES

Until recently, the preponderance of the solid-waste industry in the United States had remained in private hands. Unfortunately it was, and in some sense still is, an industry with an extremely poor image in the eyes of the general public. Many people believe that the most unsavory elements of American society control the industry and dictate its policies. It is not surprising then that it is difficult to recruit and retain labor – either in the private or the public sectors of the industry. It is not surprising that until recently the industry has been ignored by most of the leading American industrial concerns. Thus the development of new solid-waste process in the United States has been left in the hands of myriads of small, undercapitalized, and under-staffed concerns. Because of the highly complex nature of composting-reclamation, technical and market development for these processes have suffered in the extreme.

Finally, following the lead of the Europeans, the Federal Government established a vigorous solid-waste program. This program, administered by the Public Health Service, has fostered projects in all fields of solid-waste management and has been especially effective in attacking the important problems related to composting.

Pathogen Survival

One of the most important of these problems relates to a need to understand the survival rate of pathogens during aerobic digestion. As mentioned previously, the 150-ton-per-day Gainsville, Florida plant is operated by a joint arrangement among the Public Health Service, the University of Florida, and the Metropolitan Waste Conversion Corporation. One of the primary research projects at this plant is a study of pathogen survival.

The Europeans have been active in studies of pathogens for a longer period. Dr. D. Stranch of the Justus Liebig University at Giessen has made extensive use of the "inoculated-bag" technique [18]. By this method, a small bag of refuse is injected with a known number of pathogenic organisms and is placed in the larger body of refuse to be processed. As anticipated, kill times shortened considerably with increasing temperatures and increasing amounts of pre-grinding the refuse. Approximate kill times for salmonella and anthrax bacteria were five days in a Dano drum, 18-21 days in a windrow, and 24 hours in a Multi-Bactor tower, [11]. Naturally the rate of kill will vary substantially with the type of pathogen or virus being tested. It would appear that most widely used composting processes effectively dispose of pathogens if aerobic conditions and a corresponding high temperature rise are maintained in the mixture. Nevertheless, the subject deserves further study especially in light of the trend to develop more economic processes having shorter cycle times in the digester.

An alternative approach to pathogen destruction has recently been offered by Solid Waste Sciences, Inc., a company which proposes to build a composting plant in Lake County (Cleveland) Ohio. The SWS process calls for the standard steps of size reduction, partial salvage and aerobic digestion. However, as a final step, the company proposes to install a low-energy electron accelerator manufactured by High Voltage Engineering Corporation. This unit is to provide a final radiation treatment to the entire compost stream to ensure the complete destruction of all pathogens. No doubt, if the intensity of such radiation were sufficiently high a complete kill could be achieved, but it is unclear that such a process step can be justified economically.

Mechanical Handling

A further, but more elusive, problem in seeking more widespread application of composting-reclamation processes is the general difficulty encountered in handling large quantities of municipal refuse. At first thought, it would appear that the design and operation of composting facilities would be a natural and attractive diversification for the large chemical and petrochemical companies. Indeed, there are many analogies between compost production and the complex processing carried out routinely by the process industries. This fact has not gone unnoticed by many of the larger chemical companies which are constantly seeking new business opportunities and new ways to invest their enormous cash flows. Nevertheless, despite several years of very intense study, very few companies have entered the field. And those that have, appear to have lost large sums of money. One contributor to these substantial losses has been the practical difficulties encountered in merely handling municipal refuse on a day-to-day basis. Most process industries have become accustomed to handling material streams which have highly consistent properties. Obviously such consistency cannot be expected in handling municipal refuse; composition may vary substantially with changes in the season and variations in the day-to-day weather conditions. It has been this wide variation that apparently led to the difficulties encountered by some of the major companies that have sought to enter this field. Not only must the process equipment, especially conveyors and grinders, be designed for these changing conditions, but also the operating personnel must become accustomed to wide variations in operating procedures.

Future Developments

The increasing participation of industry in solid-waste management that is now occurring should eventually yield great benefits for all concerned. New types of process equipment were reviewed in Chapter 8. Other innovations are also under study which seek to fabricate new types of construction materials from raw refuse. Work in this area appears promising but is in a very primitive state; progress has been hampered by a lack of funds for research efforts.

In the final analysis of the present status of composting-reclamation, it is clear that the process continues to exist on a very infirm foundation. The professional farmer and nursery man remain unconvinced that compost can be utilized economically in agriculture or horticulture. The paper industry is not yet able to utilize the type of waste paper currently being salvaged in reclamation plants. Neither have increases been realized in the amounts of glass and various metals being recycled; the problems here appear to be the lack of purity currently achievable in the reclaimed materials.

Against this backdrop of frustration, it continues to be clear that we simply must develop means of reusing discarded materials. Today this belief is supported in fact only by a few isolated, successful operations in Europe and in theory by the qualitative belief that there must be a better way of handling our solid waste. However, only when the the qualitative belief of the layman can be converted into the economic and technical beliefs of the professional, will we see increasing use of reclamation in this country. Continued and intensified programs from the Federal government appear to be the only feasible means to this end.

REFERENCES

1. Joint working party on municipal compost, Interim Report, Surrey, England, 1959.
2. C. I. Harding, "The Nature of Compost," Brochure provided by Metropolitan Waste Conversion Corporation, Wheaton, Illinois, 1968.
3. L. Kehren, and S. Vaillant, "Methods of Household-refuse Processing and the Importance of Aerobic Composting in a Tropical Environment," Informatory Report No. 23 of the Central Bureau for the Study of Equipment for Overseas, Housing and Urbanisation Department, 1962-1963.
4. Sir Albert Howard, "An Agricultural Testament," Oxford University Press, 1940.
5. S. A. Gothard, "The World Context of Refuse Disposal," *Municipal Engineering,* Feb. 18, 1966.
6. E. Arditti, "Properties and Uses of Municipal Compost," *Municipal Engineering,* Feb. 18, 1966.
7. R. E. Johnson, "Agricultural and Horticultural Use Studies," *Municipal Engineering,* Feb. 18, 1966.
8. Joint working party on municipal composting, "Origins of Working Party," *Municipal Engineering,* Jan. 29, 1965.
9. D. Gumpert, *The Wall Street Journal,* June 23, 1970.
10. P. W. Joy, The Carborundum Company, "A Presentation to the New York Society of Security Analysts," June 26, 1966.
11. S. A. Hart, "Solid-waste Management/Composting; European Activity and American Potential," Public Health Service Public Publication No. 1826, U.S. Department of Health, Education, and Welfare, Solid-wastes Program, Cincinnati, 1968.
12. J. H. Prescott, "Composting Plant Converts Refuse into Organic Soil Conditioner," *Chemical Engineering,* Nov. 6, 1967.
13. Roto-Shredder Company, product brochure, Crestline, Ohio, 1969.
14. D. Brennan, "Composting Cuts Disposal Costs," *Metropolitan,* May 1965.
15. E. G. Spitzer, "Composting Works in Houston," *The American City,* October, 1967.
16. Johnson and Anderson, Inc., and Ducker Research Company, "Economic Feasibility Study for Michigan Refuse and Sludge Composting Plant," *Compost Science,* Autumn, 1968.
17. H. T. Blackhurst, Report to Metroganic Sales Company from Texas A & M University College of Agriculture.
18. H. J. Banse, and D. Stranch, "The Importance of Pre-fermentation in Composting," *Compost Science,* 7, 3, Autumn-Winter, 1966.

INDEX